D1593948

Bad Girls of Japan

Bad Girls of Japan

Edited by

Laura Miller

and

Jan Bardsley

BAD GIRLS OF JAPAN
© Laura Miller and Jan Bardsley, 2005.

All rights reserved. No part of this book may be used or reproduced in any manner whatsoever without written permission except in the case of brief quotations embodied in critical articles or reviews.

First published in 2005 by
PALGRAVE MACMILLAN™
175 Fifth Avenue, New York, N.Y. 10010 and
Houndmills, Basingstoke, Hampshire, England RG21 6XS
Companies and representatives throughout the world.

PALGRAVE MACMILLAN is the global academic imprint of the Palgrave Macmillan division of St. Martin's Press, LLC and of Palgrave Macmillan Ltd. Macmillan® is a registered trademark in the United States, United Kingdom and other countries. Palgrave is a registered trademark in the European Union and other countries.

ISBN 1–4039–6946–9

Library of Congress Cataloging-in-Publication Data

Bad girls of Japan / edited by Laura Miller and Jan Bardsley.
   p. cm.
Includes bibliographical references and index.
ISBN 1–4039–6946–9 (hc)
   1. Women—Japan—Social conditions. 2. Sex role—Japan. 3. Feminism—Japan. 4. Women—Psychology. I. Miller, Laura. II. Bardsley, Jan.

HQ1762.B334 2005
305.42′0952′09045—dc22   2005045966

A catalogue record for this book is available from the British Library.

Design by Newgen Imaging Systems (P) Ltd., Chennai, India.

First edition: November 2005

10 9 8 7 6 5 4 3 2 1

Printed in the United States of America.

For our colleagues and friends

Contents

Illustrations

Acknowledgments

Bad girls have been on our mind for some time. Our research on media attention to women in Japan has introduced us to numerous icons of deviancy created to capture a particular historical moment. There were the well-known New Women and Modern Girls as well as Salary Girls, the Real Estate Siren, Philandering Housewives, Hip *Bijin* (Beauties), Split-Personality Women, and that maven of the leisure ethic, the Three-Hour-Lunch Woman. Tracing the paths of these and other bad girls in Japanese popular culture and establishment codes led us to articles, presentations, edited journal issues, and now to this volume of essays. Two panel presentations were especially fruitful in stimulating our thinking and so we thank the participants from both "The Bad Girls of East Asia: Taiwanese, Koreans, Japanese, and Okinawans who Break Rules, Flout Gender Norms, and Upset People," presented at the American Anthropological Association, New Orleans, in November, 2002, and "Between the Princess and the Martyr: Spectacles of Hope and Betrayal in Postwar Japan," presented at the Association for Asian Studies in Washington D.C., in March 2002. We owe a hearty thanks to Sally A. Hastings, one of the editors-in-chief of *U.S.-Japan Women's Journal*, for encouraging us to explore this topic through editing special issues of *USJWJ*. We also thank the authors involved with "Speculating on Spin: Media Models of Women" (*USJWJ* No. 19, 2000), edited by Laura Miller, and "Women for a New Japan: Sex, Love and Politics in the Early Postwar" (*USJWJ* No. 23, 2002), edited by Jan Bardsley, for joining us in these projects.

The opportunity to produce a book on the topic of bad girls owes to the good fortune we have had to work with our Palgrave editor, Anthony Wahl. We are grateful for his continued enthusiasm, support, and eye on production schedules. Heather VanDusen, editorial assistant, has provided much-appreciated advice on the preparation of this manuscript. Lori Harris prepared the book's index with skill, making sure no bad girl went unnoticed. We also acknowledge our thanks for the publication grant from the University Research Council at the University of North Carolina at Chapel Hill that helped us complete the manuscript.

Many have graciously given permission for the use of images. Our sincere thanks to: Meiko Ando for her illustration of "Onibaba"; Kelly Foreman for her photograph of Imafuji Chōjūrō IV teaching *nagauta shamisen* to geisha; Nakahara Sohji and Himawariya for the use of the dust-cover of Yoshiya Nobuko's book, *Flower Stories*; Kyodo News Service for the use of the photo of Abe Sada; Watanabe Yayoi for the use of an illustration from her manga, *Bachelor Party*; Bungei Shunjū for the use of the dust-cover image of Nakamura Usagi's book, *Shopping Queen*; Ueto Rie and Osaka Shino for their *purikura* photos; Jeffrey Chiedo for his photograph, "Girl in a crowd"; Nobue Suzuki for her photograph, "A Filipina–Japanese couple, christening their son." We are especially pleased that the artist Hisashi Tenmyouya granted us permission to use his work, "Chiba Lo-Rider Girls" from his "Notorious Street Group" series on the cover of *Bad Girls of Japan*. We thank Kiyomi Kutsuzawa, an editor-in-chief of *U.S.-Japan Women's Journal*, for granting kind permission to include in chapter 6 some portion of Gretchen Jones' article, "'Ladies' Comics': Japan's Not-so-Underground Market in Pornography for Women," originally published in *U.S.-Japan Women's Journal English Supplement* 22 (2002): 3–31.

We extend our deepest appreciation to the band of bad girl authors who joined this project: Rebecca Copeland, Melanie Czarnecki, Kelly Foreman, Sarah Frederick, Hiroko Hirakawa, Gretchen Jones, Sharon Kinsella, Christine Marran, Katherine Mezur, and Nobue Suzuki. From start to finish, this has been a group project. Email exchanges and cards, reading and rereading each other's work, helping each other locate images and sources has produced one book, numerous good times, and enduring friendships. When Miriam Silverberg agreed to join the project, becoming the Mega Bad Girl who would write the Afterword, we were pleased indeed. We also extend warmest thanks to Rebecca Copeland, veteran editor and author, for all her support and advice, and, of course, for her invention of the secret handshake.

We met as graduate students in the 1980s at UCLA when we worked together on the graduate student publication, *Journal of Asian Culture*, and were active in the Graduate Students Association. We continue to receive encouragement from the friends we made at UCLA and from those who have come into our lives since. We thank our partners Roland Erwin and Phil Bardsley for being such a bad influence on us. Now that this project is complete, we will dedicate ourselves to becoming Three-Hour-Lunch Women.

Laura Miller and Jan Bardsley
February 2005

Introduction

Laura Miller and Jan Bardsley

In 2001, during a night of carousing, an all-girl biker gang engaged a rival girl-gang in a brawl. Skirmishes of this sort are not particularly noteworthy but this one was different. Before the score was settled one of the gangs attacked the other with Molotov cocktails.[1] Needless to say, as Molotov tossing is against the law in Japan, all the women, ranging from early teens to early twenties, were arrested. "[We'd seen] this TV show where a helmeted student hurled a Molotov cocktail at police and thought it would be a good weapon in our fights." When asked about the fight, Tanaka Tomoko, gang member and beautician, gave this explanation to the police.

Tanaka and her gang of biker chicks capture attention and prompt curiosity. What caused the beautician and the high school teens who followed her to engage in such dangerous behavior? More than a few might also wonder how *girls* could get into the kind of street brawl normally associated with boys. It is these questions and the cases of other such Bad Girls that we take up in this volume. Women who defy patriarchies, whether they are interpreted as liberatory models or serious malefactors, provoke intense concern, censure, and public debate. Visibly transgressive, they direct attention to the borders of propriety even as they threaten to alter them. Worse still, they can make deviance look fun, as if they are devils at play, as if they revel in being Bad Girls. They make the Molotov cocktail toss look like a blast. But do these Bad Girls and their subsequent fallout simply serve as cautionary archetypes and juicy fodder for scandal consumption, or might they also be inspiration for transformation?

Who decided that they are bad in the first place? In a TV interview the reggae musician Bob Marley was asked by a snide reporter, "Why is it that Rastafarians have such a bad reputation?" Bob replied that, "I wouldn't say that Rastafarians have a bad reputation, I would say that people give them a bad reputation." Looking at some of the Bad Girls in this volume, we agree with Bob. We think that many so-called Bad Girls, like their Rastafarian counterparts, are not themselves intrinsically bad, at least from our feminist perspective. Rather, their badness was attributed to them by a sexist and male-dominated society that has attempted to define, limit, and control women. It seems that especially during periods when women's roles and freedoms were and are changing, anxiety over these changes is often deflected onto a search for the Bad Girl, who is named and then analyzed for the social disruption which, it is hoped, she will contain within her "deviancy." Her libertine sexuality, her assertiveness, and her perceived disrespect are labeled "behavioral problems" thus sidetracking scrutiny away from such social issues as class inequities, tension in the family system, and the arbitrariness of patriarchal constraints on women. Earlier scholarship on the New Woman and the Modern Girl has demonstrated this.[2] More recently, as *Bad Girls of Japan* shows, the rebuke Japanese women have received for their assumed badness has bordered on the ludicrous/grown even more ludicrous. Brand-savvy women have been scolded for supporting French nuclear experiments by purchasing Chanel. Certain female photographers have been held accountable for the recent trend in making revealing and strange self-photographs, as if the stimulus for their portraits were self-aggrandizement rather than a symptom of widespread girls' discontent. Conveniently, it is all her fault. If that Bad Girl would only behave as she should, like a decent girl, the world would be right again.

Such convenient use of the Bad Girl makes us curious about what subversive potential exists when out-of-line women become the focus of public attention. *Bad Girls of Japan* explores this issue, examining the interconnections between female agency, gender ideologies, and Japanese models of womanhood. In doing so, this volume builds on recent scholarship on Japanese women by focusing on a single characteristic—badness—(rather than a particular occupation or political bent), and investigates it from multiple perspectives, cases, and historical locations. Not only does *Bad Girls of Japan* bring to light new views of gender politics in Japan, it also offers ideas and examples for making connections with transgressive women in other cultures. Is the scandal over Imelda Marcos' shoe collection related to the concern for the Japanese love of Louis Vuitton? Does performance artist Norico use shock for the same purposes as the equally politicized comedian Margaret Cho? Do the experiences of Filipinas in Japan and their Bad Girl strategies suggest comparison with the immigrant lives of women elsewhere in the world? Analyses in *Bad Girls of Japan* opens the way for questions like these and other thoughts about how the Bad Girl emerges in any popular culture, serving as entertainment, cause for concern, and as a celebrated and envied outlaw.

From destructive goddesses to degenerate schoolgirls; from penis-severing lovers to bold exhibitionists, as a representative of disorder and malfunction, the Bad Girl always points beyond established ideas of normativity and propriety. She resists easy definition and defies containment. At the same time, the Bad Girl cannot be separated from the specific time and place that gives rise to her. For these reasons, we believe that it is unproductive and even impossible to try to encompass all her incarnations in a single template. Instead, we consider the Bad Girl as an ever changing category, and one that shifts with public perception. As chapters in this volume demonstrate, the Bad Girl and the Good Girl of one era can easily trade places with each other in the next. Prewar Bad Girl writer Yoshiya Nobuko, with her flowery girl-centered works, seems downright wholesome when compared to contemporary graphic S/M ladies' comics. What once seemed so daring is now so benign. For that matter, many Japanese parents would no doubt be happy to return to the days when all schoolgirls did was meet boys in parks for the exchange of kisses.

Our task has been to find new ways to think about the diverse Bad Girls who have captured attention in Japan, and to devise our own strategies for disrupting common assumptions about them. As scholars from several disciplinary homes, we have employed varied theoretical perspectives in tracking the Bad Girl. We challenge cultural meanings of Bad Girls by asking how they are constructed, what their functions are in specific settings, and how we can think critically about the stereotypes, representations, and characterizations that have been ascribed to them. Our compass permits a degree of indeterminacy or opposition in the different chapters. For example, not everyone regards the category of geisha the same way, and descriptions and reference to geisha in this study vary widely. Similarly, scholars have their own styles of writing, and their work leads them to draw different conclusions about bad girls.

Reclaiming the Bad in the Girl

As other scholars have noted, there is a delicate difference between "men's use of women for sexual titillation from women's use of women to expose that insult."[3] There is always a risk of reinforcing the misogyny one seeks to undercut. Inevitably, our interrogation of Bad Girls must walk a fine line between critique and complicity. At the same time, we wish to scrutinize the ways in which girls and women are negatively named and described. Scholars have historically had an interest in how women are labeled, and how these labels may reflect sexist, stereotypical viewpoints.[4] A critique of such words and descriptions is important because they often serve as vehicles for assumptions that are uncritically accepted as normal, and therefore go unchallenged. A feminist perspective assumes that use of such denigrating labels and descriptions does not simply reflect an individual's nasty opinion or attitude, but is the manifestation of patriarchal social structure.[5] One goal of a feminist analysis is to explore the ways in which negative labeling and defining fortifies a patriarchal system. Throughout this volume, authors consider how Bad Girls were named and defamed by others, and how women have reclaimed Bad Girl names and statuses for themselves.

Like all use of language, the term "girl" is dependent on context for its meaning. In English, the word "girl" is often used to belittle and infantilize adult women in mainstream society. Inspired by lesbian feminists who preempted deprecatory labels by making them terms of empowerment, others began to reclaim once derogatory labels and markers for women and the things associated with them. There is also recent resurrection of the term "lady," rejected as taboo by many older feminists but now used with heavy irony in feminist and lesbian zines, music, and art contexts. In her analysis of the term "grrrl," Rachel Fudge describes how a concept originating in a girl-centered post punk subculture changed from denoting "girlpower" to a defanged and bland synonym for girl or woman.[6] She charges that conservatives in the United States have also hijacked "girl" and "grrrl" to enhance their image. Our reclamation of the English word "girl" in this volume is a feminist one, used by feminists spanning different ages and backgrounds, as an ironic in-your-face statement of female agency. Although some chapters in the volume focus on the Japanese concept of the *shōjo*, we are not equating these two terms, but rather use "girl" in an entirely English language and feminist manner.

A recent trend in English-language feminist scholarship is to explicitly problematize the Bad Girl. The American flapper, popular music divas, and historical figures such as Mae West are reevaluated by scholars interested in understanding the ways in which their threatening or contentious personas and performances thwart social assumptions and expectations.[7] Similarly, our inquiries in this volume take us through films, magazines, books, manga, conduct manuals, and other popular media that carry Bad Girl cultural constructions in Japan. By proposing alternative views of Bad Girls, we hope to discern the social and historical formations which have established and promoted gendered norms. We aim to shift thinking about Bad Girls from viewing them as necessarily morally deficient or pathological, to seeing them as gendered subjects in patriarchal structures. As Lila Abu-Lughod said "Respect everyday resistance not just by arguing for the dignity or heroism of the resistors but by letting their practices teach us about the complex interworkings of historically changing structures of power."[8]

We cannot ignore, however, the potential risk involved in naming this book *Bad Girls of Japan*. The title calls to mind a precocious naughtiness that threatens to trivialize or demean women in Japan. Many of the Bad Girl images as well point toward the pornographic and the comic. At a time when popular fiction such as the novel *Memoirs of a Geisha* or the film *The Last Samurai* are proving exceptionally influential in shaping interest in Japanese culture abroad, a book on Bad Girls also runs the danger of functioning as one more case of marketing "weird" or "exotic" or "sex paradise" Japan to an English-language audience. Are we guilty of sensationalizing the transgressive woman in the same way for which we attack the Japanese press and conservative critics in the book's chapters? Is our case different because we tend to respect her deviance rather than denounce it?

The Bad Girl's deviant choices, whether willfully determined or, as Marley's comment suggests, *projected* upon her, are entwined with sensationalist journalism. Her choice to careen outside the boundaries of proper behavior necessarily renders her exotic, outside the norm. It is she who revels in undoing any attempt to transcend the politics of the trivial, the nonsensical, and the shocking, for all that is bad about such labels is integral to her persona. There is simply no *good* way to talk about Bad Girls. At the same time, we realize that blaming the bad girl lets everyone else off the hook, and we would be remiss not to consider this process, too, as part of the bad girl phenomenon. Certainly, we believe that the Bad Girl is often the unwilling product of her society. Within the classic sociological formulation, social problems should be understood as rooted in social structure, and are not based simply on individual failures or shortcomings. Yet "blame-the-victim" discourses are common in popular media representations of social ills such as poverty, violence, and sexual harassment. Thus we also find that many times the Bad Girl is the target of individual blame and is held solely responsible for her own problems. In her research on prostitutes who were opposed to the Prostitution Prevention Law of 1956, G.G. Rowley found that they were supposed to be rehabilitated into low-wage good-girl jobs by virtue of the law. However, although they didn't like being prostitutes, they didn't want to be consigned to poverty either.[9]

If anything, the last twenty-five years of scholarship in English on women in Japan has celebrated women's resistance. Whether they are the heroic foremothers, who pioneered political and consciousness-raising change such as Fukuda Hideko, Kishida Toshiko, and Ichikawa Fusae, or today's covertly subversive office ladies, smart-mouthed aerobics instructors, crime novelists, or zealous community activist-housewives, these women, scholars argue, are nobody's dupe.[10] Scholars also maintain that the preference for conceptualizing all Japanese women as the docile and obedient "geisha girl," a view well entrenched in American popular culture, says more about American anxieties over femininity than reflects a knowledge of Japan. As a means of correcting this perception, the Massachusetts Department of Education's *Guidelines for the Bias Review*, for example, stipulate that textbooks must not introduce "Japanese women who are servile and obedient." Guidelines from publishers Macmillan, McGraw Hill, and Holt, Rinehart and Winston forbid showing, "Modern Japanese women wearing kimonos or carrying babies on their backs."[11]

The Bad Girls in this volume cannot be accused of being servile or obedient, and the contemporary ones neither wear kimono nor even have babies to carry on their back (and if they did, the backpack would be a snappy, custom-made one). But it would be a mistake to leap to the conclusion that these Bad Girl images constitute more progressive portraits of women in Japan. Replacing one image with another does little to challenge monolithic characterizations. Moreover, we maintain that it is not the image per se that is the issue. Rather it is the environment in which images are deployed, circulated, and interpreted that gives them tangibility and power. Tracing the life and the potency of Bad Girl images is one of the goals of *Bad Girls of Japan*. Also, we must keep in mind that even potentially disruptive images may be neutralized once they are incorporated into mainstream media. Dick Hebdige furnishes a useful analogy with his analysis of how British subcultures are returned to the symbolic social order through a continual process of recuperation.[12] The recuperation in the media is done primarily through two strategies. One is to trivialize or naturalize the behavior. To extend this to our Bad Girls, this is done by making fun of them through rhetorical moves that dismiss them as laughable objects of ridicule. Another method to defuse their threat to society is to place them beyond the pale, to exoticize them through labeling them as deviant and aberrant. Thus the challenges posed by Bad Girls are met and defused by extreme media redefinition that neutralizes them. Because of this process of media recuperation, Bad Girl efforts at resisting do not always create new alternatives as much as new models and categories that are similarly conscripting.

We must also consider whether resistance always merits approval. Among the Bad Girls in this volume are real and fictional women who suffer from their choices. Murderer Abe Sada does jail time. Self-proclaimed shopping queen Nakamura Usagi reports that her endless consumption ultimately leaves her feeling empty. Hagiwara Hatsuno, the lead character in the Meiji novel *Demon Winds Love Winds*, pays for her ambitions by betraying her best friend, becoming guilt-ridden and finally dying. Is the system set up so that the Bad Girl's resistance burns as no more than a temporary flame of freedom and passion? Would she be better off working within the system rather than traversing its borders? These are also questions to contemplate in thinking about the Bad Girls in this volume and in reception of them.

What does this study tell us about Japan? According to the journalist Jiro Adachi, Japanese woman are so constrained by good girl proscriptions that in order to dodge traditional ways and morality they have to flee toward supposedly less homogeneous, repressive, and sexist "Meccas" like New York.[13] Here they are not unlike the imaginary geisha-girl Sayuri in Golden's novel, who unable to find "liberty" in stodgy Japan, somehow could only find "freedom" in New York City. Perhaps the American ethnocentrism of these journalists and authors prevents them from recognizing a type of leniency that could only be possible in Japan. In a society so circumscribed, where small infractions may be taboo-why not go all out and be really "bad?" Yet Bad Girls can and do get away with things in Japan, and their transgressions incrementally allow for a shift in norms. They often have big, devouring appetites—for sex, for adventure, for freedom, for handbags. Bad Girls are important to Japanese society, they provide extreme examples that expand the continuum, opening up space for what is acceptable. Taking them seriously takes us into new places, new perspectives, new paradigms.

Regardless of the differences in the nature of badness we examine, one problem we try to address is the way in which Bad Girl categories and understandings deny women other potential identities. The geisha is never given her due as the connoisseur of a music genre or the lifelong member of a socio-artistic guild, she is rather a sexualized object of male attention. The Filipina is never an individual with her own unique dreams and desires, but is rather a poor bride of Japan or a mercenary sex-worker. These one-dimensional representations strip individual women of humanity. Although ascriptions of women as bad are often organized in opposition to notions of propriety and correctness, this is a false dichotomy that conceals the reality of individual lived experience. The putatively unvirtuous schoolgirls, masochistic heroines, wanton goddesses, unrepentant brand shoppers, and graffiti writers we examine are uniformly funneled into negative conceptual slots, thus obscuring the diverse and complex ways in which they represent a struggle for autonomy and self-definition.

The cast of transgressive women that we have chosen to present in this volume comprises a continuum of Bad Girls, all of whom have deliberately disrupted Japanese society in some fashion, whether or not they intended their actions to be read as bad. It is the cosmically bad who set the stage: the fearsome women of ancient Japanese legend: the hideous goddess Izanami, the voracious mountain witch (yamamba), and the shape-shifting Dōjōji maiden. These are followed by geisha past and present, whose brazen choices separate them from both good, domestic women in Japan and "geisha girl" fantasies in the West. Gliding onto the scene next, we have the degenerate schoolgirl of the early 1900s astride her Dayton bicycle, eager to fulfill her modern ambitions and girlish dreams. These dreams are given flower when our next Bad Girl, Yoshiya Nobuko strides up to the podium, enchanting her audience with whimsical tales of budding love and secret girl spaces. Yoshiya's blossoms give way to the notorious Abe Sada, whose 1936 crime of killing her lover and severing his penis is explained in multiple and changing ways, casting doubt about whether she is bad or good. Then, it is time to open the curtain on a contemporary cast of Bad Girls—those who create and consume the sexually explicit, violently graphic ladies' comics; shoppers gone wild over luxury brands, teenage girls who craft obscene, satirical self-photos, and teenagers who shock with their extremely dark make-up. When Filipinas in Japan take the Bad Girl stage next, we see how the meanings of bad and good change in light of transnational experiences. Our final Bad Girl, performance artist Sunayama Norico, literally claims the limelight in locales all over the world, giving new meaning to the fun and play a Bad Girl can express, and the trouble that she can cause.

Degrees of badness expressed here vary widely and no doubt readers will find themselves reacting differently to each Bad Girl. Abe Sada's crime can make us squeamish, bad schoolgirls can be endearing, obscenity can offend. Norico is hell bent on shocking. One reader's idea of a genuinely misguided woman might be another's model of break-out independence. We propose that in fact, it is the Bad Girl's capacity to provoke varying, even vehement reactions that plays such a large part in her rise to prominence, and that compels us to watch her. She dares us to react, to interpret, and to come to terms with her choices.

In selecting our Bad Girl cast, we chose those who stood out as challenging dominant codes of proper feminine behavior in different eras and in often different ways. By considering each of these Bad Girls in turn, readers will not only learn why they were viewed as deviant, but also understand how their transgressions revealed the borders of goodness, borders often devised to keep all girls in their place. Thinking about what these Bad Girls, even those of widely different times, have in common, one can detect certain themes that make connections among them.

What Makes a Bad Girl?

What is it about good and bad behavior that draws people to make lists? Sei Shōnagon, the acerbic Heian court lady, did it, speaking frankly about what she found appealing, distressing, and hateful about the people around her. For example, her list of Things That Don't Have Any Redeeming Qualities includes "Ugly people with disagreeable personalities."[14] The crabby Neo-Confucian scholar of Tokugawa times, Kaibara Ekken, author of *Onna daigaku* (Greater Learning for Women) compiled quite a long list of behaviors that women had to guard against, including talking overmuch and giving into silliness.[15] Modernity brought new lists. Conduct manuals for schoolgirls sold well in the early 1900s, and in 1928, members of a roundtable discussion elaborated a list to describe the Modern Girl and what was wrong with her, citing such offenses as her use of direct language, her aggressive sexuality, and her habit of accosting men for train fare.[16] In Japan today, good girl/bad girl lists can be had in many forms, from company training manuals to how-to manuals such as *Anthology of Office Lady Taboos*.[17]

We conclude this Introduction by suggesting yet another list. A Bad Girl List that illustrates the connections we see among the chapters in this volume. While each chapter considers the Bad Girl in terms of her particular historical and cultural location, themes emerge that unite them across time as well. We should also add here that our interest in Bad Girl history is not intended to discredit scholarly depictions of heroic good mothers, chaste daughters and hardworking wives, or the women who might have adopted these titles with pride. Rather, our project intends to add to the range and panoply of possibilities that have existed for women in Japan.[18] While the historical contexts shift and change and the "degree" of badness mutates across the generations, one constant is the effect the Bad Girl has in her own time—her power to disrupt is met with an equal power to contain.

Bad Girls Are Scandalously Visible

As Maxine Hong Kingston put it in *The Woman Warrior*, "Isn't being a bad girl almost like being a boy?"[19] Often, Bad Girls are simply doing the things men do, which certainly does not sit well with everybody. The curators of Japan's national femininity, for example, have been aghast at the way women in the new millennium have been carrying on. Critics complain that girls put their make-up on in public, speak loudly on their cell phones on trains and in restaurants, and no longer take care of themselves or their rooms and clothing. Newspaper stories, letters to editors, and TV news coverage often chastise girls' supposedly deteriorating public decorum. One of the uncivil behaviors they stand accused of is "eating in the open air," a behavior thought to be particularly unfeminine and low-class. But, with the increase in fast-food restaurants and convenience stores selling ready-made meals, many girls buy foods they can eat on the go. They carry their Seven Eleven Yakisoba Beef Croquettes and Yakiniku Rice Burgers out into the street and eat while sitting on a curb or building ledge. Although "standing noodle shops" in train stations have provided convenience for commuting salarymen for decades, girls' public eating is viewed as unacceptable.

One of the features that several of the Bad Girls in this book share is that they have been operating as if they have the right to do what they like, wherever they happen to be. Their confident exploitation of space and resources, once considered a male privilege, shocks and outrages. Even good girls who tried to stay within the parameters of propriety found themselves inching nearer the Bad Girl edge, as Rebecca Copeland has noted of Meiji period women writers.[20] The Bad Girls described here often challenge male ownership of the public sphere, and conflate distinctions between private and public domains. The Meiji schoolgirl riding her bicycle through the park and the postmodern, over-tanned teenager parading down the street on ten-inch platforms both move into public areas with confidence, knowing that they will stand out. In their excursions, Bad Girls sometimes blur the distinctions between the public and the private domain. Performance artist Norico uses public space for the display of her bottom under her engulfing dress, and girls who make provocative photos send them in to be published in their favorite magazines.

Bad Girls Make Too Much Money

During the Tokugawa era, transgression of boundaries by women was a staple source of dramatic tension in Japanese vernacular narrative, especially Kabuki plays. The Tokugawa regime challenged rules of sexual behavior, mainly adultery and private prostitution. Yet by tracking female crime and punishment during this period, Diana Wright has shown that government efforts to legally define and restrict women were mainly driven by economic considerations rather than ideology.[21] In her research on the Modern Girl, Miriam Silverberg discovered that the "free-living and free-thinking" ways ascribed to the Modern Girl had much to do with her economic self-sufficiency.[22]

Similarly, many of the authors of *Bad Girls of Japan* wonder if anxiety over perceived wickedness is actually displaced resentment of female economic success and independence. Japanese men have squandered millions on status markers such as Suntory Special Reserve Whisky or George Low Wizard 60 golf clubs, yet they are not denigrated for doing so. Bad Girls, on the other hand, are chastised for earning too much money as geisha performing at parties, or for being overcompensated for writing girls' fiction like Yoshiya Nobuko, whose paychecks were fatter than the men elected to the Japanese Diet. Observers begrudge the success of brand-obsessed writer Nakamura Usagi, who made and spent a fortune writing fantasy fiction, and continues to make and spend an enormous salary as a nonfiction essayist.

Bad Girls Push Girlish Behavior to Extremes

Bad Girls push good girl habits into Bad Girl territory. Normal make-up, shopping, cute friends' photos, and flowery writing are taken to extremes. Bad Girls often use the "trappings of femininity"[23] to stage power. Through employment of a hyperfeminine mode, they mask the outward appearance of threat. Bad Girls take the denigrated markings and features of culturally defined femininity and celebrate them. The sweet and girlish elements in Yoshiya's writing, which incorporates novel orthographic devices such as exclamation points, dashes, and line breaks, adumbrates similar new writing styles seen in Bad Girl photo graffiti. The use of the pink hyper feminine often confounds observers. Is the performance artist Norico going along with gendered expectations with her over-the-top "girlness," or is her Bad Girl act creating a new kind of girl culture? Which side is she on? When the contemporary *ganguro* ("black face") layers on all that make-up, is she accommodating to the requirement that girls not appear with natural and "naked" faces, or is she spoofing it?

Bad Girls' Bodies Are Out-of-Control

Just as mythical Bad Girls have uncontrollable passions and messy bodily transformations, real ones often revel in openly crass displays of offensive behavior, raw language, and gritty explicitness. The writer Nakamura Usagi, for instance, wallows in outrageous rants about her own base desire for money and brand goods. She is not shy about describing her bouts of diarrhea or her belching. Bad Girls are depicted as so ravenous and uncontainable they are like animals, consuming all within their grasp—flesh, bone, and designer bags. Male assessors decided that Abe Sada did not have normal erotic feelings but was rather driven by an animalistic sexual drive. More recently the *ganguro* described in men's media is characterized as having low, greedy, animalistic urges. Bad Girls don't simply read racy comics, they go in for explicitly masochistic ones full of incest and other taboo subjects. In one of comic artist Watanabe Yayoi's works, a few erotic scenes are not enough, and sexual acts are found on nearly each of its 30 pages. Bad Girls don't just murder a lover, they whack off his appendage as well, or like the Dōjōji maiden, become the phallus. Bad Girl excess is notable in the work of Japanese women performance artists who, having nothing to lose by virtue of their status outside male-dominated circles, are free to produce their own forms of creative agency unfettered by any social rules whatsoever. Their mind-blowing acts transform bodies from objects into sassy subjects that disconcert the onlooker.

Girlhood Itself is a Dangerous Time

The period between childhood and adulthood is a dangerous time, fraught with potential for growing up into a good woman or growing out into a bad one. Many of our authors discuss Bad Girls who are caught in this stage of in-between-ness, which anthropologists often refer to as a state of crossing over or "liminality." This period of border-crossing and ambiguity is seen in chapters that focus on the *shōjo* and the discourses that have surrounded her. Many traditional societies have marked this period with rituals of transition that involve masquerade. Perhaps contemporary girls who make obnoxious or defaced photos use them as a type of mask or costume, turning the space of the photo into a time to play with new identities. The liminal characteristic of the Bad Girl seems enhanced when she is creating such new spaces in society, whether through the boardinghouse, the geisha house, or the shopping street.

Bad Girls Do What They Want to Do

In a magazine article asking men about bad girls, some of the women they single out as especially reprehensible are those who, while they are at a love hotel and he is in the shower, use their cell phones to call other guys, and those who check their phone messages right after they finish having sex.[24] Bad Girls in this volume likewise forge ahead in pursuing their own dreams and desires. They flout conventions for female appearance, think of their own needs and aspirations, and engage in dangerous displays of modernization. Geisha allowed their white teeth to go unblackened, modern prewar writers bobbed their hair, and postmodern girls sport dark tans and bleached tresses. The shrine maiden Okuni dared to dress in male garb and carry a Christian cross pendant and a samurai sword in her performance. Her presumptuous act was followed by others who dared to break rules, such as town geisha who illegally performed in non-sanctioned venues or donned male clothing. The mythical *yamamba* opts to live alone in the mountain.

Bad Girl nonconformity also takes the form of choosing to live a life not dictated by male needs and desires. One of the most deeply entrenched requirements for good female character is that a woman always think of others first. For decades, the worst trait that Japanese men have named when asked what they hate most in women is selfishness or self-centeredness. Of course, what is considered self-centered in women is gauged differently when it surfaces in men. Bad Girls in this volume often exhibit what male critics would consider self-centeredness: geisha who reject marriage so that they can devote themselves full time to their careers, writers who live openly with female lovers, and women who spend their pay on brand goods rather than saving it for their families.

Bad Girl Influence on Good Girls

During the 1920s, Japanese media gave sensational coverage to upper-class women who confronted or left horrid husbands or dealt with other personal traumas.[25] Female readers admired their courage, and while they recognized the gulf separating them from these illustrious personages, they nevertheless were impressed that such "ideal" women had similar problems. Common women afterward felt freer to write stories and letters revealing their own Bad Girl behaviors. This then is the downside of making an example of Bad Girls—good girls might get ideas from them. Many of the chapters in this volume make it clear that Bad Girls are dangerously alluring. Meiji schoolgirls, geisha, Abe Sada, and *ganguro* have been featured as exemplars of the weird or improper in news stories, comics, movies, fiction, editorials, and other forms of media. Yet at the same time, holding such prototypes up for censure runs the risk of boosting their glamour and appeal. As readers become aware of the variety of experiences reported about other women, they gain an enhanced and critical awareness of what they share with others, leading to experimentation and the ability to generalize experience. Other women may furtively admire these paragons of badness and begin to emulate them. Women start sending their own S/M stories in to be published, good town wives copy geisha fashion, and schoolgirls everywhere get fake tans. Eventually, these encroachments stretch the criteria for what is considered good, leading to the revision of boundaries. Bad becomes good.

In the late 1990s, female singer-song writer Aikawa Nanase wrote a song entitled "Bad Girls" (title in English) after a visit to Shibuya, where she saw a carload of party girls zoom by in a flashy vehicle. Her song celebrates girls as living in a carefree manner and rejecting self-sacrifice and docility. Aikawa herself is something of a Bad Girl who dropped out of high school at age 15, and her songs and her persona give a public face to girls who are often not part of the Japanese social narrative. With this volume, we wish to join Aikawa in bringing attention to Bad Girls of Japan, examining their contributions, their protests, and their vitality.

Following East Asian practice, all Japanese names are cited with the family name first, except in instances where the person writes in English or other non-Japanese languages. Another exception: the case of performance artist Norico who does not use her family name, finding it a bit of patriarchal baggage she'd prefer to do without.

Notes

1. "Girl Gangs Brawl with Molotov Cocktails," *Mainichi Shimbun*, June 26, 2001. Online at <http://mdn.mainichi.co.jp/news/archive/200106/26/20010626p2a00m0fp006002c.html>.
2. For analysis of the Modern Girl, see Barbara Sato, *The New Japanese Woman: Modernity, Media, and Women in Interwar Japan* (Durham, NC: Duke University Press, 2003), and Miriam Silverberg, "The Modern Girl as Militant," in *Recreating Japanese Women, 1600–1945*, ed. Gail Lee Bernstein (Berkeley: University of California Press, 1991), 239–266. For the New Woman, see Jan Bardsley, *The Bluestockings of Japan: New Women Fiction and Essays from Seitō, 1911–1916*, forthcoming from the Center for Japanese Studies (Ann Arbor, MI: University of Michigan).
3. Lucy R. Lippard, *From the Center: Essays on Women's Art* (New York: Dutton, 1976), 125.
4. In *Bitches, Bimbos, and Ballbreakers: The Guerrilla Girls' Illustrated Guide to Female Stereotypes* (Penguin Books, 2003) the Guerilla Girls deconstruct and reclaim English labels used pejoratively against women. Kittredge Cherry wrote about Japanese words that reveal cultural attitudes, which he called "instructive words," for a Japanese feminist journal. His collection was later published as a book that describes how gender-based expectations and assumptions are coded in the Japanese lexicon. Kittredge Cherry, *Womansword: What Japanese Words Say About Women* (Tokyo: Kodansha International, 1987). See also Laura Miller, "Bad girls: Representations of Unsuitable, Unfit, and Unsatisfactory Women in Magazines," *U.S.-Japan Women's Journal English Supplement* 15 (1998): 31–51.
5. A point made many years ago by Jane Flax, "Women Do Theory," *Quest* 5, no. 1 (1979): 20–26.
6. Rachel Fudge, "Grrrl: You'll be a Lady Soon," *Bitch: Feminist Response to Pop Culture* 14 (2001): 22–24.
7. Angela J. Latham, *Posing a Threat: Flappers, Chorus Girls, and Other Brazen Performers of the American 1920s* (Middletown, CT: Wesleyan University Press, 2000); Lori Burns and Melisse Lafrance, *Disruptive Divas: Feminism, Identity and Popular Music* (New York: Routledge, 2002); Pamela Robertson, *Guilty Pleasures: Feminist Camp from Mae West to Madonna* (Durham, NC: Duke University Press, 1996).
8. Lila Abu-Lughod, "The Romance of Resistance," *American Ethnologist* 17, no. 1 (1990): 53.
9. G.G. Rowley also notes how the furor over this issue was framed in terms of virtuous middle-class women wishing to rescue the fallen woman, a framework that ignored Japanese and American men's patronage of the system. G.G. Rowley, "Prostitutes Against the Prostitution Prevention Act of 1956," *U.S.-Japan Women's Journal English Supplement* 23 (2002): 39–56.
10. For scholarship on these famous foremothers, see, for example, Sharon Sievers, *Flowers in Salt: The Beginnings of Feminist Consciousness in Modern Japan* (Stanford: Stanford University Press, 1983); Gail Lee Bernstein, ed. *Recreating Japanese Women, 1600–1945* (Berkeley: University of California Press, 1991); Vera Mackie, *Feminism in Modern Japan: Citizenship, Embodiment and Sexuality* (Cambridge: Cambridge University Press, 2003). Research on contemporary Japanese women includes Laura Spielvogel, *Working Out in Japan: Shaping the Female Body in Tokyo Fitness Clubs* (Durham, NC: Duke University Press, 2003); Yuko Ogasawara, *Office Ladies and Salaried Men: Power, Gender and Work in Japanese Companies* (Berkeley: University of California Press, 1998); Amanda Seaman, *Bodies of Evidence: Women, Society, and Detective Fiction in 1990s Japan* (Honolulu: University of Hawai'i Press, 2004); Robin LeBlanc, *Bicycle*

Citizens: The Political World of the Japanese Housewife (Berkeley: University of California Press, 1999).

11. Diane Ravitch, *The Language Police: How Pressure Groups Restrict What Students Learn* (New York: Alfred A. Knopf, 2003), 191.

12. Dick Hebdige, *Subculture: The Meaning of Style* (London: Routledge, 1979), 96–99.

13. Jiro Adachi, "How Q found her groove," *The New York Times* (January 30, 2005).

14. Sei Shōnagon, *The Pillow Book of Sei Shōnagon*, trans. and ed. Ivan Morris (New York: Columbia University Press, 1991).

15. Kaibara Ekken, *Women and Wisdom of Japan*, trans. K. Hoshino (London: John Murray, 1905).

16. Silverberg, "The Modern Girl as Militant," 250.

17. Zennikkū Eigyōhonbu Kyōikukunrenbu, ed. *OL tabū shū* (Tokyo: Goma Seibo, 1991).

18. Similar to the film *Women in Japan: Memories of the Past, Dreams for the Future*, produced by Joanne Hershfield and Jan Bardsley, 2002, our chapters depict women of diverse backgrounds and occupations. *Women in Japan* was screened by the filmmakers at several venues (women's centers, community centers, colleges) in different locations in Japan (Tokyo, Sapporo, Kobe, Otsu, Kyoto, Yasu) in the summer of 2002. Audience discussion afterward revealed that the film's intentionally positive approach to the accomplishments of the six women featured drew mixed reactions. While many found the film inspiring, others believed that it was too positive a portrait and failed to show the struggles women still face in Japan.

19. Maxine Hong Kingston, *The Woman Warrior: Memoirs of a Girlhood Among Ghosts* (New York: Random House, 1977), 56.

20. Rebecca L. Copeland, *Lost Leaves: Women Writers of Meiji Japan* (Honolulu: University of Hawai'i Press, 2000), 222–223.

21. Diana E. Wright, "Female Crime and State Punishment in Early Modern Japan," *Journal of Women's History* 16, no. 3 (2004): 10–29.

22. Silverberg, "The Modern Girl as Militant," 247.

23. Debbie Stoller, "Feminists Fatale," in *The Bust Guide to the New Girl Order*, ed. Marcelle Karp and Debbie Stoller (New York: Penguin Books, 1999), 45.

24. "Dansei 100 nin no shōgenshū kawaii onna v. kawiikunai onna," *Say* no. 229 (July 2002): 115–118.

25. Sato, *The New Japanese Woman*, 110–111.

"Onibaba" by Meiko Ando (2002). Reprinted with permission from the artist.

1

Mythical Bad Girls: The Corpse, the Crone, and the Snake

Rebecca Copeland

Just as there is an archetype of woman as the object of man's eternal love, so there must be an archetype of her as the object of his eternal fear, representing, perhaps, the shadow of his own evil actions.

Enchi Fumiko, *Masks*[1]

"Faster," the woman thought to herself as she raced up the mountain path. The lights of the village winked in the distance. She could hear the temple bell tolling as night enveloped the valley. "Faster." The mountain path was steep. Here and there massive roots leapt out of the darkness. She had to be careful not to trip. Branches tore at her arms and legs and tangled her hair. "Faster." Her breath came out in short gasps, disappearing into a white mist behind her. Her heart pounded, her temples throbbed. One step more, two steps, and then it happened. Beneath her feet the roots began to undulate in gentle submission, carrying her upward. The branches brushing softly against her skin, buoyed her forward. The ache in her arms and legs melted into the night air; the tangle of hair cascaded down her back in a stream of silken threads. And then she felt it. First it rumbled in her stomach like thunder and as it made its way up through her chest and inside her throat, her body grew warm. She threw her head back, squeezed her eyes shut, and opened her mouth to let it out. A long exhilarating laugh. Running wildly through the mountains, arms outstretched, mouth pulled wide in laughter, she was ecstatic. As she slipped through the trees into a clearing on the ridge, she met the moon head on. Someone was watching. Call it a sixth sense, call it an acute sense of smell, but she knew there was a man on the other ridge. She could smell his fear as he crouched behind a rock watching. Call it a sixth sense, call it an acute sense of hearing, but she could hear what he thought. What an evil woman, running alone in the mountains. Wild, hungry, she wants what she can't have. She wants me. Call it a sixth sense, call it an acute sense of sight but she could see herself as he saw her. Hideous, naked but for her filthy hank of hair, her lips curled back to reveal a gaping emptiness. Ashamed, she slunk back under the darkness of the trees; ashamed, she crawled back into the dampness of her cave; ashamed, she slithered back into the fetid tangle of her nest; ashamed, she stepped back into her mountain hut, wondering the whole time, "Am I really evil?"

Are women inherently evil? Mythological constructions of women across many cultures would have us believe so.[2] Chinese myth represents woman as *yin*—all that is dark, illogical, and wet. Her negativity must be moderated by the positive masculine principle of *yang*. In Bali, where the cosmos struggles between good and evil, the female is cast in the role of the latter—personified by the frightful witch Rangda. Greek mythology offers up Gorgons, whose female faces are so hideous that any who behold them turn to stone. Biblical traditions invest woman with Original Sin—sensuous, beguiling, and treacherous, she tempts men toward damnation. Hindu mythology complicates the equation with Kali—death-in-life. A womb that devours, she gives life and then takes it back. Japanese myth too inscribes woman with the unknown, the uncontrollable, that which invites desire and inspires dread. As fecund as she is destructive, she is all that exceeds the limits of imagination and control.

Images of female power are plentiful in Japanese myth and range from the bright awesomeness of the Sun Goddess Amaterasu to the dark frightfulness of the *yamamba* or mountain crone. Multi-layered and interconnected, it is impossible to explore the complexities and richness of *all* the mythological bad girls in a study of this length. For the sake of argument, therefore, we examine three mythical figures in this chapter, the corpse, the crone, and the snake. In our investigation, we explore the way Japanese myths and succeeding narratives have marked the female form with mystery, fear, and desire. We will see how the male privilege to name and control has acted against original female power. Threatened by the potential of female strength to disrupt the male-defined social order, mythmakers rescripted female potency and consigned it to positions of abjection and debility. Izanami, the primal female in Japan's early myth narratives, is powerful and beneficent in her fertility, but she speaks out of turn, usurping male authority. And so she is removed from the realm of words. Made to inhabit the depthless, cavernous world of Yomi, the domain of death, she becomes the very embodiment of both desire and defilement. Another of her ilk is the ravenous *yamamba*, or mountain crone. Perspicacious, invested with magical powers, the mountain crone is able to transgress the limits of human reason. But her distinctive taste for human flesh—particularly male flesh—makes her loathsome; an odious outcast. Finally, we discuss the Dōjōji maiden who, beautiful yet unpredictable, transforms from gentle maiden into snake—an image of sexual energy and doom. Whereas each mythological construction offers a beguiling tale, the result of these female images exceeds mere entertainment. The dangerous female is contained and controlled by the power of language, or logos—embedded as she is in cautionary tales that predict and warn of her evil.

The Corpse

"Don't look," Izanami demanded of her consort Izanagi before leaving him standing outside her chamber door in the underworld darkness of Yomi. The primal female deity, Izanami had been mortally wounded giving birth to fire. Not content to lose his beloved spouse, Izanagi had gone in pursuit of her and finding her in the treacherous regions of the underworld, bade her return to him. Izanami was not at leisure to cross the boundaries between worlds easily. She needed permission. Izanagi waited outside his spouse's chamber door while she consulted with the others. He waited and waited and when he could wait no more, he pulled a tooth from the comb in his hair and, lighting it as a torch, pushed his way into the forbidden chamber. What he confronted was horrifying. His beautiful spouse stretched before him little more than a rotting corpse, her decomposing body squirming with maggots. "In her head was the Great Thunder, in her breast was the Fire Thunder, in her belly was the Black Thunder, in her private parts was the Cleaving Thunder, in her left hand was the Young Thunder, in her right hand was the Earth Thunder, in her left foot was the Rumbling Thunder, in her right foot was the Couchant Thunder; altogether eight Thunder deities had been born and dwelled there." Terrified, Izanagi turned on his heels and fled—infuriating his spouse who pursued him along with a cohort of hags. In an effort to foil their pursuit, Izanagi brandished his sword—to no avail. In desperation, he snatched the headdress from his brow and flung it down. Immediately it turned into grapes, which the hags happily gobbled up—but all too quickly. They continued to gain on the male deity. He threw down his comb, which turned to bamboo shoots—beguiling the hungry hags long enough for Izanagi to make for the opening of the underworld cave. At the entrance he found peaches, which he hurled like cannon balls at the hags, sending them reeling in retreat. When Izanami herself rushed forward to nab her husband, he flung a giant stone into the opening of the cave, thus sealing permanently the passage between the worlds of life and death.[3]

Myths open themselves to varied approaches and interpretations, and this particular sequence, paraphrased from the eighth-century *Kojiki*, or *Record of Ancient Matters* (712), is no exception. It has fueled the inquiries of religious studies theorists, anthropologists, archeologists, and psychoanalysts among others, and has seeped into the collective imaginations of subsequent myths and stories. What is notable for the purposes of this discussion is the way the Yomi underworld is signified by the female deity that resides at its center. The dark, cavernous, chaotic spaces of this mythic realm is not unlike the female body itself—with its hidden chambers, its oozing, and unpredictable changes. Sealed within this female-like space, the body of the female deity becomes the site of death and defilement. She becomes a corporeal map of awe and horror. Beset with maggots, her once beautiful form transforms into a swelling, chaotic rumble, her head, breasts, belly, and private parts delineated by thunder.

Ancient worldviews frequently equated the female with the impure, often with evil itself. Given that her body was the site of discharges and emissions, of miraculous change and transformations, she has been suspect of harboring all that is dangerous and threatening. Her power was most apparent during her menses or at childbirth, when her body demonstrated the amazing capacity to reproduce itself. Women in these states were thus labeled unclean and defiled. The labeling was in itself sufficient to yoke women to feelings of shame. But to insure women would not be allowed to disrupt social order while in these "defiled" states, patriarchal societies enforced taboos and ancillary purity rites that demanded their temporary removal. In extreme cases women were isolated in parturition huts for designated periods of time. Practices of parturition redoubled belief in female impurity and inspired notions of female duplicity. The non-menstruating, non-lactating female may appear outwardly gentle and pure, but because her body contains the capacity for defilement so great that it merits ostracism, the female is surely implicitly and permanently corrupt.[4]

But why must the female be corrupt? Why not the male? The male form, with its seemingly uncontrollable erections and ejaculations, ought to be as easily represented as mysterious and monstrous. Certainly male bodies decay as readily as female. Then, in the Japanese myth cited above, why is the primal female deity inscribed as the locus for Yomi, the world of death, and not her male consort? Perhaps this is because men have been in charge of telling the story—writing the histories, the religious doctrine, the Law.

The alignment of females with evil in Japan was the result of what must have been a long and protracted process. A trace of this process may be discerned in an early section of the *Kojiki*. When the primal pair are first dispatched to the land below to begin the process of creating the islands of Japan and the life on it, Izanagi and Izanami engage in a brief courtship. They circle around the pillar of heaven, each heading in a different direction, and when they meet, Izanami, the female, exclaims: "How delightful! I have met with a lovely youth." Izanagi is displeased that Izanami has spoken first, so he has them undertake the process again. This time when they meet on the opposite side of the pillar, he exclaims in an echo of the female's earlier statement, "How delightful! I have met a lovely maiden!"[5]

Once again, this particular myth sequence is open to many interpretations. For the purposes of this discussion, however, it would seem to suggest a silencing of female prerogative and to point to usurpation of female power. Little is known about the social and political structures of Japan prior to these historical records, and what little *is* known is the subject of intense debate. Nevertheless, significant evidence suggests that a shamanic queen named Pimiko ruled an early federation on the Japanese islands. According to the Chinese sources that provide the account, she "occupied herself with magic and sorcery."[6] Unmarried, she kept one thousand women attendants and only one man in her service. He was responsible for bringing her food and drink and serving as her medium of communication with the outside world. The location or even veracity of this early federation is hotly contested. But archeological evidence throughout Japan supports the presence of an early shamanic tradition in which women were—if not paramount—clearly significant.

Another example of early female shamanic strength is that of Empress Jingū (c. 169–269 CE), who appears in the early Japanese histories as the consort of Emperor Chūai and the mother of Emperor Ōjin. This semi-legendary figure, whom some suggest is another representation of the above-mentioned Pimiko, is believed to have had extraordinary ritual powers, powers that allowed her to transmit the wishes and expectations of the deities. In keeping with one such deified directive, Jingū is said to have launched an attack on neighboring Korea.[7] The image of the warrior queen, girded for battle, also conjures forth the spectacle of Amaterasu, the Sun Goddess, who in mythological accounts waged a winning battle against her brother, Susa-no-wo, the Storm God. Angry with her brother's willful, wailing petulance, Amaterasu slung a giant quiver full of arrows over her back, wound her hair into bunches, and strode out purposefully to meet her brother—so aggressive in her anger that with each stride she sank into the earth up to her thighs.[8] Once she quelled him, Susa-no-wo retreated to the earth where he became the patriarch to many. The Goddess of the Sun remained in her heavenly abode—magnificent and luminous—the mother to emperors.[9]

Carmen Blacker, in her brilliant study on shamanistic practices in Japan, notes that Japan originally followed a Siberian type of shamanism where "sacral power was believed to reside more easily and properly with women, and where in consequence women were recognized to be the natural intermediaries between the two worlds."[10] The physical characteristics of the female body—its ability to transform, to engulf, to invite—that would be the source of its pollution were also the basis of its power. Women did not merely serve as passive conductors for spiritual possession, but actively initiated, maintained, and performed sacred ritual. Pimiko, whoever she might have been, surely held "sacral power." But by the time the *Kojiki* was compiled, this era of female prominence was fading.

Buddhism, which entered Japan in the sixth century, hastened the usurpation of female power by diverting attention away from female strength—with all its dangerous, disorderly temptations—and placing it on female pollution.[11] In the process, the new religion positioned rigorously trained male ascetics at the center of religious ritual. "The Buddhist figure of the ascetic, in appropriating to himself the active, banishing side of shamanic practice," Blacker notes, "relegated the task of medium to one of mere passive, almost automatic utterance."[12] The once powerful female, "the true natural shaman"[13] became a mere conduit for spirit transmission.

Buddhism, with its various taxonomies of enlightenment and suffering, saw women as the very embodiment of all that bound humanity to the wheels of fate. Believed to be more closely linked to nature and its instinctual impulses, women epitomized carnality, impurity, and excess. According to Elizabeth Wilson, "because women bear children, they represent . . . the illusion of reproductive immortality, the folly of continued life in an impermanent world ravished by death."[14] There is much debate over the purported misogyny of the Buddha's teachings; after all, transcendence of gender is one of his important tenets. But regardless of original intent, the manner in which societies practiced Buddhism across the Asian continent and in Japan reflected those societies' misogynistic impulses.

When Buddhism entered Japan along with other continental influences, it offered new political opportunities for those who could claim an investment in it. The exercise of this political clout required that the existing political and sacral powers be disenfranchised. Gradually, therefore, Buddhist adherents marginalized those who had protected and promoted Shintō ritual. The dwindling of indigenous powers paralleled the denigration of female strength. Again, to cite from Blacker:

> Under the new regime the appearance of mantic queens such as Pimiko and the Empress Jing[ū] became impossible. Under the new system too the *miko* in the large shrines began to lose her mantic gift, and to become before long the figure she is today. Decorative in her red trousers and silver crown, she now dances, sings, assists in ritual but no longer prophesies. The mantic power with which this ancient sibyl was endowed passed from the large shrines to the level of what Robert Redfield called the "little tradition," the largely unrecorded, orally transmitted folk religion of the villages.[15]

The newly evolving male-identified prerogative minimized female power—and the political threat it suggested—by shunting it to the periphery. Female strength was deprived of its potency by either being reduced to positions of degradation or exalted to positions of such inviolable splendor they were beyond the reach of most mere mortals. To an extent, the lofty deification of Amaterasu accommodated the divestment of original female power. Placing her in counterpoint to Izanami's darkness, Amaterasu became all that was pure, mighty, and inaccessible to the disenfranchised mortal woman with her natural bleedings and defilements. Women were terrified by Izanami at one end and taunted by the impossibility of Amaterasu at the other.

The recording of the *Kojiki* as well reveals this diminishing of pre-Buddhist authority and participation. Prior to the advent of writing in Japan, professional reciters (*kataribe*), perpetuated official records and family histories by committing all to memory and passing their knowledge to their heirs. A hereditary position, these reciters, some scholars believe, were frequently female.[16] Yet the adaptation of a written language from Chinese, eventually displaced them and shifted the transmission of knowledge to male scribes. This shift abrogated female power and subsumed the female voice under the male prerogative of *logos*, or the authority to name, order, and control.

Much as the female shaman was deprived of her active role, the female deity, Izanami, was silenced, her words appropriated nearly verbatim by her male partner. Denied power over the discourses used to represent her, woman came to stand most frequently for all that was inchoate, wordless, and "messy." Her cavernous body—dark and engulfing—housed the underworld with its circular mysteries of death and terror. Confined though she was to a subterranean world, even so the female principle refused containment or contentment. The excesses of her body—with its secretions and flows—constantly threaten to leak beyond its borders and endanger the stability of the masculine order.

The Crone

Perhaps no image signifies the danger of the uncontainable, ravenous female as readily as the *yamamba* or mountain crone. Ostracized and feared, chased to the fringes of civilized society for her seemingly insatiable desires, the *yamamba* represents a renegade power that must be controlled. A compellingly complicated figure, the *yamamba* appears in a variety of legends and myths often representing contradictory attributes. Occasionally she materializes as a positive force; transforming herself into either a comely woman or an unseen power, she swoops into villages to assist with chores. More often she is malign—wrinkled and ragged, her hair a tangle of grey, she lurks in the mountains where she preys on the unfortunates who lose their way. Her victims almost always are male.

> "Don't look," the old woman commanded the priests as she closed the door of her chamber and stepped outside to gather firewood. She would prepare them a bite to eat—meager though the meal would be. She was a poor but kindly old woman, living alone in the mountains. The priests, on their way to Kumano, had sought refuge in her hut as night fell around them and they stumbled in the darkness. They had nowhere else to go. They waited and waited and when they could bear the suspense no longer, they peeked inside the forbidden chamber. The spectacle that befell their eyes was almost more than they could comprehend. The room was piled high with corpses in various stages of decay, skeletons of men. Was this what became of the unfortunate wanderers who happened into her hut? Were they to become her next meal? Terrified, the priests turned on their heels and fled, nearly colliding with the old woman. She pursued them, her gentle countenance now transformed into that of a hideous ogress. The priests rubbed their rosaries with all their might and, with Buddha's assistance, subdued the ogress.[17]

The origin of these mountain crone stories is unknown. Scholars have offered various interpretations. Renowned folklorist, Yanagida Kunio, for example, suggests that in rural villages young women who were afraid of revealing an illegitimate pregnancy would slip off into the mountains to have their babies. Driven either by an instinctual need to flee the prying eyes of the community, or forced beyond the borders of the village so as not to pollute it with the blood of childbirth, these women sought refuge in the mountain wilds. Some women never returned, leading to conjecture that their bodies had merged with the mountain depths. And in the collective imagination of the village, the women became renegade spirits—dangerous, ravenous, haunting the borders of society.

Other scholars have drawn parallels between the mountain crone and the "abandoned granny," referring to legends in which old women who had outlived their usefulness to the village were left on mountaintops to die. Scholars disagree as to whether or not the legend has roots in actual practice. Nevertheless, notable in all these tales of women and mountains is the terrifying image of a woman alone, unattended and thereby not contained by or confined to the family structure, left to roam the fringes of society. The horror of the mountain crone shares much in common with Western images of witches, old women in forest cottages, and the "crazy cat lady"—a mainstay of any small town imagination.

In Japan, the story of the mountain crone circulated in various redactions as fairy tale, didactic anecdote, and dramatic performance. Each redaction reveals a number of features in common with the earlier Izanami myth cycle. First is the male on a quest, in search of either romance or spiritual growth. Along the way the man inevitably becomes lost, feeling disoriented as he wanders through a dark and chaotic world until he finally makes his way to a woman's chamber. She pronounces a prohibition which, when ignored by the male visitor, provokes the woman to reveal her "true" and terrifying form in all its repulsiveness. This revelation is followed by the male's flight and his fear of being devoured by the raging female. Finally, the male subdues the female. In the *Kojiki* myth, Izanagi seals Izanami in her world of death, preventing her from engulfing him and the rest of the world in her dark embrace. In the *yamamba* episode recounted above, the ogress is vanquished by the power of Buddhism—her horrific desire dissolved. In other redactions the crone "is killed by being stoned or made to eat stones,"[18] her cavernous mouth sealed forever.

The *yamamba* continues the trope of horror initiated in Izanami's cave by personifying the engulfing nature of female physiology. In another variant in the cycle of *yamamba* tales, the mountain crone appears as the perfect wife. Comely, obedient, and nurturing, she possesses all the attributes a husband might desire. Most importantly she does not eat. Or so it would seem. One night when the husband spies upon the woman, he finds that she has a giant mouth on the top of her head. As he looks on in disbelief she pulls back the hair on her head and begins to swallow enormous quantities of food. The mouth at one end of the body can easily be connected to the mouth at the other end—neither of which, it would seem, is easily satisfied. Woman is, like the cave that represents her, an orifice from end-to-end, mindless, ravenous, deadly. Meera Viswanathan observes of the *yamamba* image:

> The figure of a man-eating female demon is peculiar neither to Japan nor to premodern narratives; we see, for example, in the myths of the New World people known as the Tainos, a similar female demon graphically described as "*la vagina dentada*" (the vagina with teeth), reminding us as well of D. H. Lawrence's character Bertha Coutts in *Lady Chatterley's Lover*, with her dreaded "beak" down there that shreds and tears at men. . . . The delineation of these ravenous figures suggests an overarching preoccupation with the danger posed by female consumption as well as the need to defuse the threat, leading us to question whether the provenance of such man-eaters, ironically, is rather in the realm of male anxieties about castration than simply in female notions of resistance.[19]

The power the *yamamba* exudes, and the subsequent anxiety she induces, is not limited to her sexuality—for the male is not initially drawn to her out of sexual attraction. The *yamamba* represents all that lies outside the social norm, beyond the boundaries of the civilized. She is a woman without a family, a woman who does not conform. Cast out from the security of social sanctuary—she runs through the mountains. Her freedom figured as terror. She turns the world upside down, inverting expectations and the comfort of the assumed. In various versions of the tale, day becomes night in the *yamamba* world. The nurturer becomes the murderer. The woman who eats nothing becomes the woman who eats all.

That the *yamamba* is linked to mountains is significant. Mountains have traditionally been tied in Asian and Middle Eastern thought with spirituality, the site of pilgrimage, the signifier of enlightenment. In Japan mountains have also served as retreats from the stable and orderly world of the village, or court, or community. Romantics slipped away to the mountains to escape the prying eyes and stuffy strictures of society. Home to the subterranean world of Yomi, mountains represent that which is unbounded, dangerous, and always just beyond. In many ways in the mountains, and in the female body inscribed upon it, spirituality and the sexuality become entangled, inverted, and doubly enticing. All the fears and desires of humankind are projected on the female form, and she comes to embody the seductive desire for the uncivilized, liberated, uncontained, the yawning abyss of nothingness. This desire must be named and thereby contained and so it is imagined a hungry, horrid woman.

Once a symbol of potency and spiritual authority, phallogocentric power structures reconstituted woman's biological uniqueness as that which is abject and feared. And woman became the convenient scapegoat for not simply a generalized evil, but an evil that man felt most intimately *in himself*.[20] Taught to believe they are weaker, susceptible, biologically predisposed to insatiable desires and uncontrollable urges, women have little recourse but to accept the "universal" logic of their depraved state. Learned behavior begets archetypal patterns. And archetypes become natural law. In other words, because women have been constantly exposed to stories and images that define the female as uncontrollable, shameful, and defiled, they unconsciously accept this definition as irredeemably true. Without even being aware of their doing so, they began to locate in themselves that which conforms to this constructed definition of femaleness. And so the archetype of female evil becomes a self-perpetuating prophesy.

Feminists have long resisted the constellation of archetypes for this very reason. Not only do archetypes perpetuate either/or categories that contain women in binaries of good/bad; male/female; conscious/unconscious, but they tend to elide the distinction between woman as reality and woman as concept. That is, a reliance on archetypes does not take into account the fact that the creation and perpetuation of archetypes was largely the outcome of a male perspective, a perspective that failed to explore the difference between the male psyche that has invented female difference and the actual experience of the female.[21] Woman loses her determinability and becomes the archetype that contains her. Archetypes, it is important to note, are premised—not on actual nature but on nature *perceived*. And it is the male prerogative to perceive. As Susan Aguiar observes: "[T]he characteristics that a male believes he perceives in a woman, whether positive or negative, might not be a realistic interpretation; he might merely be projecting his own fears and insecurities or his unexamined acceptance of his own anima onto the actual female persona."[22] The female persona, the archetype of female nature, thus becomes a repository for the containment of the male viewer's own self-loathing and fears.

Containing women within archetypes of evil was one measure by which female power was controlled. Conditioning women to accede to a self-image of shame and loathing was another. The mountain crone, we can assume, was at one time an ordinary woman who lived in the wilds of nature where she gave free reign to her own natural inclinations. But because women, when left to their own devices, threaten the social order that would have them silent and submissive, the crone's "natural inclinations" are necessarily untoward, violent, and horrific. What better way to describe such antisocial passions than to render them cannibalistic? Her selfishness consumes the very fabric of social order as surely as the crone tears at the flesh of men. Japanese scholar Baba Akiko, who is highly regarded for her studies of Japanese demons, or *oni*, suggests that the mountain crone, whereas initially a woman living alone in the hills, "became an Oni because of the sudden womanly feeling of shame that overcame her when her bedroom, replete with pus and blood was exposed to view."[23] Because this exposure is the result of a man's refusal to respect her injunction, the woman is doubly humiliated. "[B]y this cruel final betrayal, by the discovery of that most secret place where she had hidden the hoard of sacrificial offerings to her passions, she was driven to the depths of shame and turned into an Oni."[24]

Notable here is the way shame is imputed to the female. Driven into the mountains because of who she is, the female must nevertheless find her condition mortifying. Unable to transcend her passions and physicality or the powers they offer, the female is taught to deny what is inherently hers and keep it hidden away. The hidden chamber, the forbidden room operates as both a metaphor for the female's strength and her sexuality. Invisible, not manifest on the exterior of the body, the female sex is hidden. The site of bloody discharge and transformations, when rendered visible would have seemed horrific.

In myths and legends, the female is forced into duplicitous roles. To the outside eye she may appear wifely, motherly—a partner to fecundity, a shelter in the wilderness. But her gentle façade merely masks her "true nature." Her true nature, her inherent evil appears as the body of the decaying corpse. Male viewers are served up visions of former beauties now displayed as putrefying carcasses,[25] or of kindly old women revealed as hideous flesh eaters, their chambers strewn with bones and pus-bloated cadavers. In these myths, the dangers typically hidden deep within the female body become outwardly visible when woman takes her "true form" and the prying eyes of the curious male expose her secrets. Transgressing the female's injunctions, the male peers into the forbidden chamber, and in seeing what he should not see, awakens from the spell cast over him. And just in the nick of time.

The Snake

The *yamamba* is easy to fear, but not so a lovely young woman. Or is she? The tale of Dōjōji reveals that if a wretched hag harbors the promise of evil, a beautiful female would prove more vicious. Of the three stories presented here, the Dōjōji tale is the most complex. Like that of the *yamamba*, it exists as folk tale, didactic anecdote, and dramatic performance. Its variants are richly complicated and so diverse that they defy easy summation. The following presents only a portion of the Dōjōji tale.

> "Look at me," the young woman commanded the priest as she crawled into his bed. She had agreed to give lodging to him and an older priest on their way to Kumano. "[F]rom the time I first saw you, this afternoon, I have longed to make you my spouse," the woman whispered in the priest's ear. "My husband is dead and I am a widow. Take pity on me." Although the woman was attractive, the priest was resolute. He was on a holy pilgrimage and could not break his vow of celibacy. But the woman was so persistent that finally he promised her that he would fulfill her desires on his return trip. She left him in peace, and he and the older priest quit her house the following morning.
>
> The woman waited for his return. She waited and waited and when she could wait no longer she realized that she had been tricked. He had no intention of returning. No intention of fulfilling her desires. Infuriated she locked herself in her room where soon she died. Her maidservants wailed with despair until they saw a forty-foot long snake slither out of her chamber and make straightaway to the road where the Kumano pilgrims passed. News of the event spread like wildfire, eventually catching up to the two priests. "Undoubtedly, because the promise to her was broken, the mistress of that house let evil passions arise within her heart and became a poisonous snake and pursues us." The priests hastened to Dōjōji Temple. There the clerics lowered the temple bell, over the young priest, concealing him within. A fatal mistake.
>
> Finding the temple the snake coiled herself tightly around the bell. And then, with tears of blood streaming from her eyes, she flailed the dragon head at the top of the bell with her tail again and again until it burst into flames. When the snake had slithered away, and when the flames had died down, the monks hoisted the bell, only to find a wisp of ash where the young priest had been.
>
> Years later a venerable and holy monk had a dream in which a snake came to him claiming to be the former priest. "The evil woman became a poisonous snake; in the end, I was made her captive and became her husband. I have been reborn in this vile, filthy body and suffer measureless torment." The former priest beseeched the monk to copy a chapter of the Lotus Sutra and dedicate it to him, so that he might be redeemed. When the monk does so, he dreams again and sees the priest and the woman, restored to their human forms, and reporting to have been reborn in Paradise. All who heard of these events were amazed by the pure and sacred power of the Lotus Sutra.

"You see, therefore, the strength of evil in the female heart. It is for this reason that the Buddha strictly forbids approaching women. Know this, and avoid them."[26]

Variants differ as to the identity of the miscreant female. In some versions she is the innocent daughter of the local innkeeper. Teased by her father into believing a traveling priest plans to marry her, she grows enraged when he does not and pursues him, growing more monstrous as she goes. Infuriated by the denial of her desire, she gradually takes on serpentine features. When she reaches the Hidaka River, the ferryman refuses to grant her passage. Undeterred, she dives into the water and emerges a fully transformed fire-breathing serpent. Now nothing can stand in the way of her desire. The priest who would shun her, cowering beneath the bell, is doomed. Wrapping him tight in the dark womb-like chamber of the bell she roasts him alive with her serpentine desire.

The Dōjōji tale exemplifies the dangers men face when interacting with women. Whereas her winsome figure may beguile and intoxicate, her charms only mask her hideous "true nature." The quality of the female nature to transform and mutate is frequently related to nature—to natural phenomena and to creatures of nature. In this regard, equating the female with the snake emphasizes the mutability of her physical self. The snake seasonally sloughs off its skin just as women's bodies slough off blood at regular intervals. Its body engorged is not unlike that of the pregnant female. So, too, does the image resemble a penis. In East Asia, the gendering of the snake is notably fluid, sometimes female, sometimes male. The instability of the snake's gender identification has served to render the image even more potent. Carmen Blacker writes of this fluidity:

[W]e shall find ourselves dogged by the figure of the snake: the snake who is transformed into a woman, who becomes a woman, the snake who forcibly claims a woman as his bride, seizes and "possesses" her, at times even kills her. We shall discover pointers relating this snake woman with our *miko* [or shrine maiden]; indicating that the shamanistic cult of which she was the central figure was closely connected with this world of water, serpents and ugly but miraculous children. Such a cult has long since been overlaid, however, by other and stronger beliefs, that in seeing in the snake woman a dim reminder of the ancient *miko* we are trying to revive a pattern of shamanist practice which has been lost.[27]

Indeed, by tying the snake irrevocably to the female, the Dōjōji tale and later myth constructions denied the image its earlier ambiguity and strength and created it as an object to be controlled and destroyed. In the Noh play version of the Dōjōji tale, for example, the snake woman is overcome by the efficacy of Buddhist prayer and rather than being redeemed is left to return to the pool of horror and desire—only to emerge again and again as a specter haunting male privilege. Woman, like the indigenous practices she represents, is neutralized by Buddhist dogma.

Associating woman with the snake—having the woman turn *into* the snake, makes of her and her body a phallic symbol. "Since women have no external organs of reproduction indicative of sexual arousal comparable to male genitals, it follows that virtually the entire female body is apt to be regarded erotically," Elizabeth Wilson notes of the treatment of women in the post-Ashokan Buddhism of India.[28] Aligning the female body with the phallus is not to suggest that woman then possesses the power associated with the phallic order. On the contrary, woman becomes the object possessed, the object that marks its possessor (invariably male) as powerful. Or, to cite Susan Klein on the subject of Dōjōji imagery:

> [T]he woman's body, the embodiment of lust, is transformed into a living phallus. . . . [W]e can see the masculine desire being projected onto the female body, a projection that enabled men to deny those negative aspects of their own sexual nature which had to be eliminated for enlightenment to occur: the woman as female snake (that is, simultaneously phallic and female) embodies the animal nature of both masculine and feminine sexuality. The "pure" monk is a passive victim of feminine passion: the danger of sexuality associated exclusively with the feminine is reinforced by the moral . . . which implies that it is the woman's blind passion alone that causes the tragedy.[29]

Woman who had earlier been made to contain all of man's fears about his own inabilities to control his passions has now become the very vessel of all man's passions. As a phallus—*his* phallus—she is vulnerable to his control. Not only does she signify female desire but she comes to contain *his* desire for her as well. As long as he can deflect his own fleshly longings and weaknesses onto her; as long as he can make *her* carry his sins, he remains pure. His ability to control the phallus by refuting the one associated with it redeems him while at the same time it condemns her to the status of sexuality incarnate.

The Dōjōji myth in its various redactions is largely didactic. Taking its cue from the warning the historical Buddha gave his disciple Ānanda to avoid women at all costs, Dōjōji similarly cautions against the dangers of the female. As Klein describes Dōjōji's explicit message: "Women are psychologically and biologically determined to a weakness of will that keeps them from being able to control their passions; women by their very existence are an inevitable obstacle to men's spiritual progress. They must therefore be avoided in daily life and excluded from participation with men in the soteriological path to enlightenment."[30]

Not only do the myths that we have discussed serve to warn hapless men of the dangers that women harbor; they also have a salutary effect on other women. Beware the monstrous serpent lurking within, women are implored. Avoid the snake-like desire that twists and coils about the heart, inflaming desire and selfish longings. See what becomes of women who give in to their own greedy passions? They become horrific, demonic, images of revulsion and repugnance. Grasping, lustful, never satisfied, they are despised for all eternity. Taught to hate their natural bodies, trained to deny their instinctive desires, the corpse, the crone, and the snake woman have for centuries served to keep female strength in check.

Postscript

Stepping out from behind the darkness of the curtain she enters the lighted space and begins to dance, slowly at first; then faster. Her hair slips from its ties and cascades around her, shimmering under the bright stage lamps. She turns, swirls, her arms released from the tension of their delicate posture begin to flail. She stamps and the stage resounds. She stamps and the mountains move. She stamps and the thunder gods take note. She stamps and all the pent-up energy of a thousand generations surges through her body from her ankles to her knees to her thighs. Kicking aside the hem of her robe she grabs her breasts in both hands and shouts, "All sleeping women will now awake and move!"[31]

Since the early twentieth century Japanese women artists have struggled to tear aside "the curtains of mystery" that have obscured the positive imagery associated with the powerful mythic women of the past. Courageous feminists like Hiratsuka Raichō (1886–1971) challenged women to reclaim their hidden sun. Whereas not referring specifically to Amaterasu, Raichō's impassioned rhetoric restored the goddess and other mythic images of productive female energy to a new generation of women.[32] Later women writers, fascinated by this trace of lost or dormant female power would invent female characters who reach back into the mythic past in an effort to reprise the mantic female. Enchi Fumiko (1905–1986), for example, presents chilling tales of female sorcery, notably *Masks* (Onnamen, 1958; trans. 1983) and *A Tale of False Fortunes* (Namamiko monogatari, 1965; trans. 2000). Writers of a more recent generation have availed themselves of the female demonic, drawing a new and positive power from the formerly abject image of the *yamamba*. Kurahashi Yumiko (b. 1935), Ohba Minako (b. 1930), and Tsushima Yūko (b. 1947), for example, return their protagonists to the mountains—often in dreamscapes—where they unearth the suppressed memory of the supernatural woman who resides there.

And so the mountains move, the rippling undulations felt across seas and continents. With a near universal appeal, the promise of hidden female potential speaks to women of various generations and cultures and manifests itself in various art forms. This chapter opens with the woodblock print titled "Onibaba"—or demon hag—designed by Canadian artist and dancer Meiko Ando.[33] It closes in homage to Ando and her reprisal of the *onibaba*'s quest for an unfettered *jouissance*. Ando was inspired by Ohba Minako's (1976) "The Smile of the Mountain Witch" (Yamamba no bishō) to choreograph the dance performance "Onibaba," which she has presented on stages in Canada and Mexico in 2000 and 2002. Whereas Ohba discloses the agony of a woman who must suppress her true powers in order to conform to social expectations, Ando allows the woman to slip free of her tethers and dance. As Rebecca Todd notes in her review of the performance "Ando transforms herself from a tightly bound village girl into the wild-haired demon who dances in [a] magical forest of projected light and shadow. In the end, the conventional young woman becomes a supernatural being. Which is, says Ando, 'what she always was in her nature'."[34]

Notes

1. Enchi Fumiko, *Masks*, trans. Juliet Winters Carpenter (New York: Random House, 1983), 57.
2. Examples of positive female strength are ample in the cultures in question as well as in others not listed here, such as the African Bantu. In Japan, as in other cultures dominated by a phallic order, female strength is often limned by conditions that make it inaccessible to most women. Either it becomes idealized and placed beyond the reach of mortal woman, or it is minimized by making it humorous, crude, or shameful.
3. Robert Borgen and Marian Ury, "Readable Japanese Mythology: Selections from *Nihon shoki* and *Kojiki,*" *Journal of the Association of Teachers of Japanese* 24, no. 1 (April 1991): 67. See also Ryūsaku Tsunoda et al. *Sources of Japanese Tradition*, vol. 1 (New York: Columbia University Press, 1958).
4. Nel Noddings discusses the invention of taboo as a way to protect men from the dichotomous nature of women—who are at once beneficent and evil. See Nel Noddings, *Women* and *Evil* (Berkeley: University of California Press, 1989).
5. Tsunoda, *Sources of Japanese Tradition*, 26. In other versions of the story the couple repeats the process after giving birth to deformed offspring and decides—in some versions on their own and in others after consulting the other gods—that the deformities are caused by the fact that the female spoke first.
6. Ibid., 6. It should be noted that the pronunciation of this name is a point of dispute and uncertainty. The characters for the name may have been read Himiko, a pronunciation now preferred. But because the sources cited herein use "Pimiko," I will follow suit for the sake of consistency.
7. There is no historical data to suggest such a battle ever took place. Some historians of Korea, in fact, suggest that Jingū was actually a princess from an early Korean kingdom who led a force of horse riders to subjugate Japan. The Japanese account of Jingū also describes how the empress began her voyage to Korea while heavily pregnant with the child who would become Emperor Ōjin. When the birth of the child appeared imminent and at an imminently inconvenient time, Jingū spoke to the child in her womb and convinced it to wait. The child waited for three years.
8. Carmen Blacker, citing the work of Saigō Nobutsuna, suggests that the image depicted in this myth sequence is not of Amaterasu herself but of a shaman, or *miko*, in the state of possession by the goddess. See, Blacker, *The Catalpa Bow: A Study of Shamanistic Practices in Japan* (London: Allen & Unwin, 1975), 105.
9. For more on the positive imagery associated with powerful mythic women, see Michiko Yuasa, "Women in Shinto: Images Remembered," in *Religion and Women*, ed. Arvind Sharma (Albany, NY: State University of New York Press, 1994), 93–119.
10. Blacker, 28.
11. For more on the way Buddhism disenfranchised women, see Junko Minamoto, "Buddhist Attitudes: A Woman's Perspective," in *Studies on the Impact of Religious Teachings on Women*, ed. Jeann Becher (Philadelphia: Trinity Press International, 1991) 159–171.
12. Blacker, *The Catalpa Bow*, 139.
13. Ibid.
14. Elizabeth Wilson, "The Female Body as a Source of Horror and Insight in Post-Ashokan Indian Buddhism," in *Religious Reflections on the Human Body*, ed. Jane Marie Law (Bloomington: Indiana University Press, 1995), 78.

15. Blacker, *The Catalpa Bow*, 30.

16. Michiko Y. Aoki, "Empress Jingū: The Shamaness Ruler," in *Heroic with Grace: Legendary Women of Japan*, ed. Chieko Irie Mulhern (New York: M.E. Sharpe, Inc., 1991), 24.

17. Accounts of the *yamamba* are found in folktales and in tales for children. They appear in the *Konjaku monogatari* (Tales of Times Now Past, ca. 1120) and other collections of *setsuwa*, or anecdotal literature. There are also Noh play accounts of the *yamamba*, such as the fifteenth-century *Yamamba*, whose authorship is unknown, and *Adachigahara* (The Adachi Moor) which is also known as *Kurozuka* (Black Tomb) and *Itokuri* (Spinning Thread). The paraphrase in this chapter is from *Adachigahara*.

18. Hayao Kawai, *The Japanese Psyche: Major Motifs in the Fairy Tales of Japan*. Translated by Hayao Kawai and Sachiko Reece (Woodstock, CT: Spring Publications, Inc., 1996).

19. Meera Viswanathan, "In Pursuit of Yamamba: The Question of Female Resistance," in *The Woman's Hand: Gender and Theory in Japanese Women's Writing*, ed. Paul Gordon Schalow and Janet A. Walker (Stanford: Stanford University Press, 1996), 242–243.

20. For more on the idea of woman as "scapegoat" see Noddings, *Women and Evil*, 36–37.

21. Sarah Appleton Aguiar, *The Bitch is Back: Wicked Women in Literature* (Carbondale and Edwardsville: Southern Illinois University Press, 2001), 17.

22. Ibid., 16.

23. As cited in Kawai, *The Japanese Psyche*, 24.

24. Ibid.

25. Whereas the reference here is to Izanami, observing the decomposing body of a former beauty was, in fact, an ascetic practice in post-Ashokan India. In an effort to awaken themselves to the falsehoods of their illusions—and particularly to quell any potential sexual desires—Buddhist mendicants would meditate on the putrefying corpses of courtesans or other renowned beauties. See Wilson, "The Female Body," 76–99.

26. Much of the information for this tale was drawn from Marian Ury, trans. *Tales of Times Now Past: Sixty-Two Stories from a Medieval Japanese Collection* (Ann Arbor, MI: Center for Japanese Studies, 1979), 93–96. The portions in quotations are direct citations. Another source for the Dōjōji tale is the later Noh play by that name. A translation by Donald Keene is available in *Twenty Plays of the Noh Theater*, ed. Donald Keene (New York: Columbia University Press, 1970), 238–263. For more on the textual tradition of the Dōjōji legend, see Susan Klein, "Woman as Serpent: The Demonic Feminine in the Noh Play Dōjōji," in *Religious Reflections on the Human Body*, ed. Jane Marie Law (Bloomington: Indiana University Press, 1995), 100–136.

27. Blacker, *The Catalpa Bow*, 78.

28. Wilson, "The Female Body," 78.

29. Klein, "Women as Serpent," 114–115.

30. Ibid., 107.

31. The line "All sleeping women will now awake and move!" is taken directly from Yosano Akiko's rousing poem "The Day When Mountains Moved," which appeared in the inaugural issue of the feminist journal *Seitō* (Bluestocking), September 1, 1911. The paragraph preceding this line is a composite of ideas inspired by Hiratsuka Raichō's impassioned rhetoric in her essay for this inaugural issue "Genshi, josei wa taiyō de atta" (Original woman was the sun) and my imaginings of Meiko Ando's performance of "Onibaba."

32. For more on Hiratsuka Raichō and this inaugural issue of *Seitō*, refer to Jan Bardsley, *The Bluestockings of Japan: New Women Fiction and Essays from Seitō, 1911–1916* (Ann Arbor, MI: Center for Japanese Studies, 2005). For more on Yosano Akiko (1878–1942), refer to Janine Beichman, *Embracing the Firebird: Yosano Akiko and the Rebirth of the Female Voice in Modern Japanese Poetry* (Honolulu: University of Hawai'i Press, 2002).

33. The author of this chapter wishes to thank Meiko Ando, a Toronto-based dancer and choreographer, and Judith Sandiford for their help acquiring this information and permission to use the Onibaba woodblock print.

34. Rebecca Todd, "A Butoh-full Mind: Meiko Ando dances a Japanese legend," *Eye Weekly*, an online journal, April 4, 2002: <http://www.eye.net/eye/issue/issue_04.04.02/arts/onibaba.html>.

The *iemoto* Imafuji Chōjūrō IV teaching *nagauta shamisen* to geisha (*geiko*) and *maiko* in Kyoto, 1999.

2

Bad Girls Confined: Okuni, Geisha, and the Negotiation of Female Performance Space

Kelly Foreman

The Geisha in Japan

The modern Japanese geisha is a versatile performing artist who derives her identity within a context that the public casts in a bad light. The widespread and deep misunderstandings about geisha result from the lack of direct contact most people have had with geisha and the subsequent reliance upon fictional and word-of-mouth representations. Geisha reside within the world of Japanese traditional performance arts, and geisha are rarely understood within this context because this world is largely unknown to outsiders. However, this context is essential if one is to understand their reasons for giving evening performances, the high costs of attending these parties, and the motivations for these women choosing the geisha career in the first place.

Both actual geisha and "geisha girls" of the West have been cast as "bad" in their respective societies for very different reasons. In Japan, the avoidance of family responsibility and monogamy to focus on careers in the performing arts renders geisha as "bad" within a society that measures female "goodness" on humility and loyalty. Moreover, because the term "geisha" was often misapplied to maids or prostitutes having nothing whatsoever to do with art, these "geisha" stories—as bought-and-sold powerless objects—are embraced as remnants of a "bad" feudal social order and incompatible with progressive modern gender roles. In the West, the "geisha-girl" image is utilized as a symbol of repression, passivity, and the inequities that we continue to find unpurged from our own society—"bad" elements that reveal stagnancy in our own social progress.

In order to untangle the various "bad girl" interpretations of geisha, we need to first gain a more complete understanding of these women and this requires us to revisit the social context of Japanese female performance history (the context in which geisha originated). Because so few have seen geisha or their performances and because the term "geisha" has been applied in a rather haphazard manner in both Japan and in the West, it is also necessary to understand how geisha are defined by the traditional arts world and to overview the essentials of the geisha as performing artist.

Geisha are defined as women who study classical Japanese music and dance, perform music and dance for parties in order to pay for their art lessons and elaborate public stage performances, and are registered officially with a central office.[1] *Geisha* is a specific title distinct from the various other women who previously worked alongside them, including *yūjo* (courtesans), *shōgi* (prostitutes), or *jochū* (maids). Moreover, while "geisha" translates as "art person," *gei* cannot be applied as in English in a general sense, such as "social arts" or "sensual arts;" *gei* refers to traditional dance, music, and handicraft arts such as pottery. Geisha devote the majority of their time to the study and performance of dance, the three-stringed *shamisen* lute and its many genres, *hayashi* (three drums and flute), and voice. Geisha support themselves and their arts study by giving evening performances called *ozashiki*, they usually retire late in life, and they prefer neither to marry nor (usually) have children.

Most of the genres of dance and music that geisha study (such as *nagauta*, *kiyomoto*, and *tokiwazu*) were designed to be performed within the Kabuki theater, and these genres constitute the focus of training even though they are often too elaborate for the small *ozashiki* teahouse setting. Geisha study these genres within guilds called *ryū*, which are socio-artistic groups led by a headmaster (called *iemoto*) and populated by students who are Kabuki performers, professional stage musicians and dancers, and amateurs. Careers in the traditional arts are very limited now, and therefore most performers in the *ryū* teach in order to make their living. Male students have the option of obtaining professional theater positions (in the Kabuki and Bunraku theaters) and salaries, but these theater positions are closed to women. Moreover, while teaching is the key source of financial support for female performers, geisha prefer to avoid teaching in favor of financial support from individual patrons. The high fees that patrons pay for *ozashiki* performances thus enable geisha to fully participate in the expensive performance arts without having to supplement their incomes with teaching.

This patronage is crucial because the costs geisha face for their full-time training in several arts genres is tremendous. For example, a kimono worn to perform an auspicious dance role can cost tens of thousands of dollars to rent, and astronomical costs for renting a theater (including scenery, musicians, dressers, wigs experts, makeup staff, backstage personnel, etc.) as well as fees to *iemoto* must also be paid. Geisha make no profit from the large stage performances they give because they are on their own to finance both performance costs and maintenance of theaters and rehearsal studios.[2] Geisha also must pay for rehearsal kimono and accessories (costing thousands of dollars), musical instruments (costing tens of thousands of dollars), monthly lesson fees (costing hundreds of dollars), fees to the teachers for participation in large recitals and concerts, and stage kimono and costumes ($10,000 to $100,000).

Because *ozashiki* are the means of covering these expenses, they are necessarily expensive. Many of the *ozashiki* patrons are connoisseurs of traditional music and dance, who study music themselves and perform with geisha during recitals and evening gatherings. Geisha do not view themselves as entertainers or playthings for men because these gatherings are done in order to pay for arts involvement, and these gatherings do not include sex.[3]

The "Geisha-Girls" We Love to Hate

The geisha described above bear no resemblance to the "geisha-girls" of the West, and the rampant misinterpretations responsible for much of geisha "badness" stem from overlooking the fact that these women are residents of the world of traditional dance and music. For example, the emblematic white face paint often described to be some sort of "badge" of the geisha is, in fact, simply a stage convention; all dancers—male or female, whether on the Kabuki stage or in a teahouse—wear this makeup when performing. However, the following description of this makeup portrays an interpretation given by an outsider:

> When a *maiko* enters a party, there is a gasp, no matter how jaded the company. She is preposterously cute. The chalk-white makeup, crimson bee-stung lips, long sway-ing *obi* and swinging sleeves give her a goofy but irresistible appeal . . . Her face [the geisha], smoothed to eggshell whiteness, becomes a blank screen onto which desires and fantasies may be projected. She is beautiful but anonymous. All traces of her uniqueness have been erased. Her eyes and mouth are highlighted and emboldened, beacons to the opposite sex.[4]

The National Geographic journalist who wrote the above passage projected her own erotic fantasy onto what she saw, and embraced the appearance as a grotesque alien beauty ideal designed for men. To her, a child wearing such makeup and costume is incongruous, "goofy." A look at this appearance from the perspective of the performing arts world reveals that *maiko* wear this thick makeup and costume because they are considered to be dancers-in-training. Dancers of the Bolshoi ballet, seen backstage in characteristic face paint and exaggeratedly made-up eyes and lips would appear similarly bizarre. The makeup is designed for audiences seated at a distance from the stage, and we are not used to seeing performers up close (the saying "a ballerina should only been seen up close by her mother" was created for a reason!). Few Kyoto customers would interpret the *maiko*'s appear-ance as "goofy;" the gasp communicated appreciation for the seriousness of this young dancer. Furthermore, most geisha do not even wear this makeup at all during *ozashiki*, unless they are performing dance.

The "geisha" image outside of Japan has been crafted and cultivated for over a century. It has been applied with great dexterity in American culture, in popular songs of the early 1920s and 1930s, in detective novels and other short stories, and in popular culture. In a 1970s cookbook, the term "geisha" was even equated with servant and cook, something actual geisha and their customers would find amusing because many possess no domestic knowledge or skills whatsoever:

> A few years ago, one of my more enterprising friends shipped a Japanese Geisha from Japan to his wife in the States. As a result, he was the envy of his associates and the

talk of the neighborhood. Even his wife liked the Geisha. And why not? The Geisha cooked and graciously served exciting Japanese food . . . When the Geisha went home, the wife carried on in true Geisha fashion. She continued cooking and serving Japanese dishes to the delight of her family and friends. And that's what this book is all about. You, too, can be a Geisha in your own home . . . So grab your kimono, light the hibachi and be the first instant Geisha on your street.[5]

While the above statement was published more than thirty years ago, more recent representations show little improvement. Arthur Golden's wildly popular novel *Memoirs of a Geisha*[6] was highly insulting to the geisha who had allowed him to interview her, and she has responded to Golden's story in no uncertain terms:

> What is written in Arthur Golden's book is false . . . He got it wrong . . . For me, personally, this is a libel, an infringement . . . also a libel against Gion as a whole . . . Real geisha don't tie men's shoes—maids do that. Real geisha don't take [time] off from their training. Golden got the organization of the geisha house wrong, and misunderstands the painted smile of the traditional *noh* dancer . . . The book is all about sex. He wrote that book on the theme of women selling their bodies. It was not that way at all.[7]

This geisha also published her actual autobiography, *Geisha, A Life*, which has gone largely unnoticed both because marketing efforts have been nowhere near those devoted to Golden's *Memoirs* but also because her account has none of the salaciousness of Golden's tale.[8] Even though Golden's story is fiction, it has been embraced globally as a true geisha biography and an authoritative account of the *karyūkai* (the geisha world); I have even encountered people who have read both and maintain that Iwasaki's story is fabricated and Golden's depicts the "real story."

What Golden's compelling novel does accurately, however, is to present a succinct example of the link between American ideas of female "badness" and the term "geisha" by spinning the tale of a pathetic slave girl, who is sold and forced to become skillful in the erotic arts. Beginning with Puccini's *Madame Butterfly*, the "geisha" idea was tied to female submission, servitude, and availability, because Cio-Cio-san (described as a "geisha") is not a performing artist within an active urban arts community of women, but is instead a "kept" woman in isolation. The foreign "geisha" idea shifted away from its original Japanese meaning due to several factors. Early Meiji-era American officials probably saw actual geisha, but would have had no aesthetic tools with which to understand the music and dance they saw during banquets at which they were guests, and assumed that the function of such independent unmarried women must be as sexual playthings. There was some justification for this assumption, because earthquakes, war, and bad economic conditions pushed many women into prostitution, and many such women were called "geisha" by unknowing customers (including foreigners) who had not had occasion to see legitimate performers with the high price tag that came with such parties. Furthermore, the eager dedication to Westernization during that time resulted in traditional institutions which differed from the West being seen as old-fashioned and "inappropriate" to the new direction of modernization. Geisha were often viewed in this light, especially because they performed Japanese arts and not Western arts (such as the piano or ballroom dance), and increasingly became an embarrassment to the new and modernized national identity.

Following World War II, the United States heard much about "geishas" and "geisha-girls." The American Occupation (1945–1952) and anti-prostitution law of 1956 triggered a shift in meaning for "geisha" once and for all away from performing arts, to mean a woman devoted to sexual services. Unlike prostitutes, geisha were able to continue working after the 1956 law was enacted because selling sex was not officially part of their cache, and prostitutes often referred to themselves "geisha" to avoid arrest. These "geisha" were numerous because there was good money to be gleaned from foreign servicemen stationed in Japan.

> In Kyoto, the word geisha smacks of the wrong type of article. It is not forgotten that [foreign] soldiers of the Allied Occupation Forces after the war used to refer to the kimonoed camp-followers and masseuses and club waitresses as "geisha girls."[9]
>
> Within a matter of months after American occupation forces arrived in Japan in the summer of 1945, thousands of the troops and officers were supporting girls as full-fledged mistresses. By 1949 it was estimated that over 80 percent of the Occupation force had at least a part-time mistress . . . "Some [Olympic] athletes insisted that they be provided with geisha for bed partners. I understand that street girls dressed in kimono were finally made available to them."[10]

Ignorant American servicemen were not interested in making social distinctions and their geisha-girl stories spread quickly back home, resulting in the firm entrenchment of the "geisha-girl/as/passive/whore" idea as essential to the American understanding not only of geisha but also of Japanese women as a group (despite how inaccurate this was). Meanwhile, the majority of the Japanese public was reading about geisha in novels, whose tales failed to correct any Western misunderstandings of geisha because the novelists themselves were unable to afford geisha gatherings and resorted to imagination instead.[11]

With some basic background knowledge about the world of traditional music and dance, the many misunderstandings about geisha should have been easily cleared up, but these "geisha girl" images remain with stubborn persistence in North America and they continue to be resurrected and reused. "Geisha nursing" means catering to male doctors; "geisha porn" is Asian women in bondage; "geisha glam" fashion refers to thong shoes, long sleeves, and cinched waists; "geisha chocolates" are soft bonbons with hazelnut filling in a package depicting a kneeling Japanese woman and Japanese flag; "geisha" in popular music means docility and apathy.[12] Countless examples such as these tell us a great deal more about gender and power struggles in the West than they do about who Japanese geisha are or where they fit in Japanese society.

I suggest that the strength and continuity of the "geisha-girl" image stems from its usefulness in defining American gender roles, that "geisha-girls" serve as a vibrant source of female "other." Joyce Zonana identifies images of non-Western women evoked in the context of feminist trope as a kind of orientalism, what she refers to as feminist orientalism.[13] The existence of non-Western institutions of bought-and-sold, pitiful women such as the supposed "geisha-girls", have the potential to assist comparative analyses of "modern" and "enlightened" women, and these images cannot avoid clouding any Western understanding of actual geisha.

The 1629 Ban on Women's Performance and the
Emergence of Geisha

If geisha were and are not the downtrodden women of lore, then who were they historically, who are they now, and what role has and does art play in their lives? The over 250 years of geisha history provide a rather surprising picture of independent, free-thinking "bad girls" who chose to bypass marriage and family. The original geisha evolved out of the performance traditions of the *odoriko* and the *tayū*, and their identity within both the traditional arts society as well as society at large is tied to the historical developments of these two female performance traditions, the theater, and the Confucian-style control of Edo-period urban society. Beginning around 1603, a Shintō shrine maiden (*miko*) named Okuni began to give avant-garde performances of *odori*-style dance while dressed as a man and brandishing both Christian cross pendant and samurai sword. The public called this *kabuki*, and the troupes of mostly women (but also some men) who followed Okuni were referred to as *onna kabuki* (women's Kabuki). However, the authorities viewed these tremendously popular performances to be so dangerous that they issued a nationwide ban only twenty-six years later prohibiting not only women's Kabuki but *all* public performances by women. The official pleasure quarters established in the urban centers functioned to contain female performers and prostitutes, allowing the government the ability to control audience size as well as the incomes and whereabouts of both performers and prostitutes. Female performers would have to wait almost 250 years to regain the right to perform on public stages, and by then men would have permanently supplanted women in the Kabuki and developed a tradition of female impersonation that would come to be seen as superior to anything an actual woman could portray.

These pleasure quarters featured prostitutes and courtesans of various levels, but customers also frequented these communities in order to enjoy music, dance, poetry, and other sensual pastimes. Because female theater performers were restricted to these quarters after the ban, many of the women within the pleasure quarters were very talented as musicians and dancers. The highest ranking courtesans were called *tayū*, whose name derives from the theater traditions of Noh and the puppet theater, where musicians (male and female) performing the musical narrative were and still are referred to as *tayū*. After the 1629 ban, the female *tayū* performers brought this title to the pleasure quarters where it came to mean the highest rank of courtesan. Over time, however, the glamorous *tayū* came to see musical entertainment as too lowly a task, and a separate class of performers called *geisha*—first male and later female—developed to serve the artistic function by which *tayū* themselves used to be defined. By 1761, the rank of *tayū* disappeared completely from the licensed quarters, one year prior to the first official listing of female geisha (1762).[14]

The other women who contributed to the development of geisha history were the *odoriko* (literally "young dancers") who gave illegal performances in samurai residences and other venues outside of the pleasure quarters. Despite the fact that they were breaking the law by performing outside of the licensed quarters, these performers were so widespread by the Genroku and Hōei periods (1688–1710) that Edo government officials in 1689 issued several decrees to limit their numbers. They banned both their dispatchment to samurai residences as well as any practice or rehearsal of music and dance. Such actions did little to dissuade the *odoriko* from performing, and many were arrested repeatedly and sent to the pleasure quarters to serve terms as prostitutes as their punishment.[15]

Because the 1629 ban made it difficult to study music and dance, the *odoriko* congregated in the theater districts located near the pleasure quarters in order to maximize access to teachers. In order to cut off these women from both teachers and the musical developments taking place in the theater, the government forced *odoriko* to move out of the theater districts in the early 1700s. However, they continued to maintain ties with the male theater world nevertheless.

> "*Jochūgei* [*odoriko* performing in Edo castle] secretly had artistic exchanges with Kabuki actors for official purposes [training]." By this description, we are able to understand the characteristics of female dance—it seems that they had exchange with Kabuki actors because their performances were the same as those on stage at that time.[16]

Odoriko were even ranked or compared with Kabuki actors in a *musume hyōbanki* (guidebook to women) published in Edo during the Meiwa period (1764–1772).[17] As primarily dancers, these *odoriko* were also increasingly becoming proficient in the *shamisen* and incorporating it into their dance performances, guided by what was taking place on the official male Kabuki stage.

> In *Edo meibutsu kanoko* [Featured attractions of Edo] (published in 1733) there was a picture of a young Tachibana-chō *odoriko* playing a *shamisen* without dancing ... Since even the total ban proved to be ineffective, repeated arrests were no more than a temporary deterrence.[18]

Meanwhile, *odoriko* continued performing in upper-class samurai residences and established stages on Edo pleasure boats in Naniwa-chō, Muramatsu-chō, and Tachibana-chō, which, along with Fukagawa and Yoshiwara, were the geographic origins for the geisha who would follow.[19] Incorporating popular tunes (*tōsei uta* and *hayari uta*) and *kyōgen* songs with *shamisen* accompaniment, the Tachibana-chō *odoriko* are seen to be the immediate predecessors of Edo geisha because they were the first to consistently possess skills in both *shamisen* and dance. Although the term *geisha* was initially used to refer to both male and female multidisciplinary artist/entertainers, *geisha* came to be used to refer to older female *odoriko* with significant proficiency in the multitude of different musical genres needed to accompany Kabuki dance forms for their younger colleagues.[20] Artistic exchange took place between these two contexts; genres from the teahouses were incorporated into the theater and theater pieces were performed within the teahouses.

And, like their *odoriko* predecessors, "free-lance" geisha known as *machi geisha* (town geisha) performed illegally outside of these walls as well. Evidently, control of female performance began to weaken even before official Meiji policy restoring women to the stage was instituted.

Female *tayū* performing the *sawari* section, the musical expression of a puppet play climax, became tremendously popular with city audiences by the middle of the nineteenth century. The female *tayū* did not perform these sections in the traditional setting of the theatre with puppeteers. Rather, they performed them as separate concert forms, often in teahouses, accompanied by *shamisen* but without puppets. Their continued popularity caused the government to issue a proclamation in 1841 specifically referring to the female performers in their general program of reforms . . . The edict states: "Persons who give theatrical performances in the downtown area have in recent times moved [out of the theater districts] and increased in numbers even though promulgations have been issued many times. Persons holding *onna jōruri* [female *jōruri*] meetings are committing an offense."[21]

Examples such as this and the ongoing decrees issued to try to control female performance outside of the pleasure quarters indicate that full control was never achieved. It would, however, take the Meiji Restoration to legalize large performances of women on public stages, and geisha would be the first to obtain the legal permission to do so.

As an overview, then, women began a nationwide theater craze that would later become the Kabuki, but were subsequently blocked from further involvement with the public theater traditions. The pleasure quarters became the only legal venue for female performers, and the tradition of the geisha grew out of a response to a division of labor within the pleasure quarters that spilled out into the cities as well. Clearly the authorities were not entirely successful in preventing female performance outside of the legal quarters, because women consistently broke laws and endured punishment in order to take lessons and to give performances of their own.

The "Badness" of Female Performers

The 1629 ban on public female performance drove women into alternative private performance venues and radically affected not only the involvement of women in music and dance but also the character of the Japanese theater. What was the basis for this severe dictate anyway? What was so bad about the early women's Kabuki performances, so transgressive as to inspire a comprehensive prohibition of *all* women from the stage? Many cite prostitution on the part of the actresses to be the reason for the 1629 ban, but prostitution was already a well-established and accepted feature of urban Japan at that time. A closer look at the women's Kabuki reveals a bit more about why officials saw it as threatening and dangerous:

They [audience members] squandered their fortunes, neglected to take care of their parents, and fathers with children did not mind indulging in jealous feelings. Therefore, if this had continued, there would have been no end to the harm to be done to the nation and its people, so the top Kyoto government administrator ordered the ban on Kabuki performed by *ukareme* [playful women, literally, women whose feet are off of the ground].[22]

Because they were women who sang and danced, their performance was named *kabuki*. I don't know when it started, but *yūjo* [prostitutes] began to perform Kabuki in playhouses, and imprudent upstarts loved it and saw it from box seats regardless of [social] position. They [the customers] still could not get enough and bought season tickets, and lost their entire fortunes in scandals, fights and disputes . . . This was harmful to the nation, it may have lead to the collapse of morals, and it was a root cause of all evil, so women's Kabuki was banned.[23]

In fact, the problem was not so much the purported selling of sex by actress-artists, but with audience behavior and social control. In the context of the Confucian social modeling being implemented, immorality here meant the indiscriminate mixing of people from differing social classes, social disorder, the raucous brawls and vandalism by audience members, and the distraction from orderly productive activity.

The establishment of the Tokugawa order did not happen without protest among the samurai themselves and without the appearance of opposition groups, among which the *kabuki mono* [performers] provide a most interesting occurrence—very much like the British "punk" phenomenon of the seventies and early eighties . . . The verb *kabuku*, used since the middle ages with the meaning of "to slant," or "to tilt," had acquired, by the beginning of the Tokugawa era, a slang usage for any anti-establishment action that defied the conventions and the proper rules of behavior. The *kabuki mono* were therefore people who expressed their anti-conformism through a series of protests against the established order, which ranged from highly unusual ways of dressing to shocking hairdos and extravagantly decorated, enormous swords . . . To the Tokugawa regime *kabuki* meant subversion and heresy, something immoral and dangerous . . . [24]

Under this Confucian model, men were expected to serve and work for the state while women were exhorted to serve the family out of the public eye. Thus, women's Kabuki needed to be abolished because these actresses and actors were not presenting exemplary examples of "good" behavior at all, as Confucian scholar Hayashi Razan (1583–1657) remarked:

The men wear women's clothing; the women wear men's clothing, cut their hair and wear a topknot, have swords at their sides, and carry purses. They sing base songs and dance vulgar dances; their lewd voices are clamorous, like the buzzing of flies and crying of cicadas. The men and women sing and dance together. This is the kabuki of today.[25]

Strict regulations existed determining who was permitted to carry weapons, but for the lower classes to carry swords and particularly for women to display them would have been scandalous. Likewise, displaying a symbol of Christianity would have also been shocking in the context of Christianity being purged and outlawed in Japan. It must have been a strange sight indeed to see both on these female performers; such actions could not have cast these women in any other light other than rebellious, dangerous, and "bad."

Bad Girls Want Fame and Fortune Instead of Family

Like the performers of women's Kabuki, geisha have been perceived throughout history as "bad," and their "badness" derives from several factors. The shirking of feminine duties to family is a key aspect of geisha "badness." Female performers in Japan have always had the tendency to transgress the norms of social order and expected behavior for women. From the Heian period *shirabyōshi* court performers to the modern Takarazuka, bad girls who concentrate on art and fame arouse suspicion and disapproval. Geisha epitomize this aspect of "bad-girl" behavior because they remain committed to this "non-female" role, potentially for their entire lives. Geisha turn away from tending families to dedicate themselves to arts teachers and the traditions of the *ryū*, and this is nowhere more symbolic than the choice to take the name of the *ryū* rather than the family name of a husband.[26] Particularly today, as Japan worries about the declining birthrate (fearing the loss of a future tax base), women who have no children are often viewed as selfish and not doing their fair share. Ironically, the same Confucian ideology that put women into the pleasure quarters in the first place and that saw the desire for fame and success as "bad" also created the possibility for the same situation to be cast as "good." Urban legends abound of filial daughters dutifully serving their parents by earning money for them as geisha, and sacrificing "normal family life" (which assumes that a woman's greatest wish is to care for a husband, children, and in-laws).

Historically, in both Edo and Kyoto, customs pertaining to both prostitutes and geisha varied widely, but it is true that some families found that a daughter with skill and grace would fare better as a geisha than as someone's daughter-in-law/wife. While such cases took place, women became geisha for a variety of reasons then and now. Today, many different types of women decide to become geisha, including college graduates, daughters of geisha, and amateur performers who lack the requisite introductions or the money for access to the high-ranking teachers of music and dance (becoming a geisha takes care of both problems). Geisha have often told me that a key motivation for choosing the geisha career is social and artistic freedom, and while it might be tempting to attribute this to modern social reforms, the following Edo-era geisha story depicts a similar situation:

> *She was in such high spirits that she must be a geisha.* A man following her is carrying a long box [her *shamisen*] wrapped in *furoshiki* looking sulky and complaining, he is saying "you're walking too clumsily."[27] [italics added]

To equate "high spirited" women with geisha certainly runs contrary to the image of the pathetic, purchased sex-slave, and it suggests that geisha were either making the best of a difficult situation or actually enjoying their exciting life as successful performers. Here the man is following the *woman* "ten paces behind."

Geisha have long been known within their communities for their strength of will, independent minds, and free-living ways, and perhaps their refusal to perform normal female roles actually contributed to their appeal.

To my surprise, the geisha-san ... were dressed as young steeplejacks enjoying a festival; these characters were all men. I was thinking that I would see lovely geisha, wearing black kimono decorated with a crest and dragging the bottom of their kimono, but I realized in the end that it was a sort of "Takarazuka." Geisha, who were career women, had as high of incomes as did men, and were allowed to have pleasant chats with prominent figures in the same *ozashiki*; and they seemed to be an embodiment of what ordinary women wanted to be. Female audience members viewed their idols playing men on stage with great envy.

The reality of geisha as a profession/career was quite far from the pathetic image of *"onsen geisha"* frequently depicted in literary works, but instead was the quintessential self-reliant career women, and [regular] women idolized them.[28]

This passage refers to a 1940s performance of the elaborate *Azuma Odori* given by Tokyo Shimbashi geisha. In the context of modern Japanese society with its myriad of choices for women, it might be difficult to appreciate the significance of women earning large salaries and bypassing marriage, but this was an unusual thing for women during this time. To describe geisha as "career women" also runs contrary to the image most have of indentured servitude.

The failure of geisha to fulfill their role as "good wives" or "wise mothers" was often most troublesome to those women humbly serving society in this way. When the Emperor Taishō assumed the throne, the Tokyo Shimbashi geisha association planned to commemorate this with a colorful parade and held a press conference to announce the details of this plan. Many women protested and tried to block this parade from taking place, with the strongest opposition coming from housewives and women's groups.[29] Two women who publicly defended and supported the geisha were poet Yosano Akiko (associated with the "New Woman" movement of the 1930s) and a progressive female educator named Yamawaki Fusako.[30]

Bad Girls Break the Law, Dress Like Men, and Are a Bad Influence on Good Town Wives

Both the *odoriko* and town geisha (*machi geisha*) flagrantly broke the law in performing outside of the pleasure quarters, toting their *shamisen* wrapped in three separate pieces to avoid being detected and arrested. Samurai residences were also breaking the law in hiring *odoriko* performers, because they were expected to avoid mixing with lower social classes. In spite of this, geisha were fashion trendsetters and influenced other urban women as well. Town geisha (performing illegally outside of the pleasure quarters) with significant *shamisen* talent were called *haori geisha* because they wore male *haori* coats to perform, brazenly ignoring official bans against this unusual fashion.[31] Donning male clothing (and using male stage names) certainly increased being taken seriously as a *shamisen* performer because the *shamisen* world is largely a male one, but such gender play was not appreciated by the keepers of moral order because respectable wives began to adopt this fashion as well.

Likewise, it was a common practice for adult married Edo-period women to shave their eyebrows and blacken their teeth, and while geisha often followed this convention as well, Bunsei-era geisha began to leave their teeth white. Since white teeth belonged to the domain of young girls, this gesture was a symbolically confusing one but it became popular and even the wives of *chōnin* (townsmen) began to sport white teeth in imitation of the geisha, to the dismay of many at the time.[32]

Bad Girls Play *Shamisen*, Good Girls Play *Koto*

From its initial emergence in Japan around 1563, the *shamisen* has always been the instrument of men.[33] Blind male performers (*hoshi*) adapted this instrument from the *biwa* tradition, and although it was women who incorporated the *shamisen* into the Kabuki during the short phase of women's Kabuki, the official music for the Kabuki (*nagauta*) has long been and is still dominated by men as performers, teachers, headmasters, and composers. The notable exception to this paradigm is geisha, who have always performed the *shamisen* within the pleasure quarters and teahouses as well as illegally outside of it. However, due to the ties between the *shamisen*, geisha, and the theater, any woman considering *shamisen* study must contend with these cultural associations. Historically therefore, "bad girls" played *shamisen* while good women—upper-class daughters and wives—did not and often still will not touch it.

In contrast, associations between women and the *koto* (21-string plucked zither) derive from the context of the Heian court, and to study it was and still is a noble pastime for women and girls from good families. This instrument has never been a feature of the musical ensembles for the plebeian theaters (Kabuki or the Bunraku puppet theater), and therefore it has few "lower-class" associations with the pleasure quarters or the theater.[34] Thus, even in modern Japan, girls from upper-class families might be encouraged to study the *koto*, violin, or piano, but they would likely be discouraged from learning *shamisen*, even though there are increasing numbers of high-ranking female (non-geisha) professional concert performers.[35]

Bad Girls Have Sex With More Than One Man

Perhaps more than any other factor, however, geisha are seen as "bad girls" because no matter how talented they may be, they are not tied to one man but may be having relationships (which may or may not include intimacy) with more than one. It is an old tired belief cross-culturally that multiple sexual partners are a man's attribute and a woman's shame, but this "badness" is compounded if the woman receives compensation in any way. When a woman is sexually involved with a man from whom she also receives money, shelter, or gifts, this woman will be viewed as prostituting if she is a geisha. If this woman happens instead to be his wife and similarly receiving from him shelter, money, or items, is she still prostituting?

The difference, many geisha pointed out to me, is that they don't have to listen to men snore, wash their socks, or cook them breakfast.

*   *   *

Dialog at a teahouse gathering with geisha, customers, and the author:

geisha 1: "Why are foreigners so interested in us anyway?"

KF: "There are a variety of reasons, but perhaps it has something to do with the fact that we don't really have a legacy of professional female performers like geisha, or a formal courtesan system."

geisha 2: "Courtesans?! We're not courtesans."

KF: "Yes, that's true, but that is how geisha are often understood in the West."

geisha 2: "You know, Kelly-chan, I hope that you're not going to write another one of those books. What's his name, that man with the story about the geisha with blue eyes?"

KF: "Arthur Golden . . . "

geisha 3: "That's him. What a strange story. But how would he know better? He's just a foreign man, and there are no geisha in America are there? How will they [Western audiences] understand your book, Kelly-chan? They like the 'Fujiyama-geisha' stories don't they? They cannot understand how we are."

KF: "My research is different. It's about what we've talked about over the years, like art and ryū, why you decide to become geisha, etc."

customer: "That will be a boring book. Nobody will buy it."

geisha 4: "You should write a book about how we are feminists!"

geisha 1: "Feminists? I'm not a feminist."

geisha 4: "Yes, we all are. We find our own way, without doing family responsibilities. Isn't that what feminists are?"

geisha 5: "But I don't dislike men . . . "

geisha 2: "That's right. We like men—we just don't do their laundry!" (everyone laughs)

customer: "You have it easy, don't you?"

geisha 4: "Do you want to play [shamisen] for the next piece?!" [art is harder than you think!]

The geisha at this ozashiki were between the ages of forty and sixty-five, so they were not young coquettes. Of course, some of the anxiety the public feels toward geisha stems from the fact that the patrons with whom geisha may be involved are often married. The pity felt for the wives stuck at home inspires contempt toward geisha and casts them as irresistible sirens that no man could possibly refuse. Those who have actually witnessed ozashiki are often surprised, however, to see that wives (as well as other women) actually join these gatherings and that the mood is usually lighthearted.

Conclusion

From the avant-garde women of the early Kabuki, to the women who broke laws against performing outside of the pleasure quarters, to the modern context in which women continue to struggle to improve opportunities in the performing arts, female performance in Japan has been a contentious issue since the mid-1600s. The public Kabuki and Bunraku theaters became all-male spaces, leaving women with the private domain of the pleasure quarters (and later the concert stage) as their performance venues.[36] Subsequently, the arts of the theater came to be seen as a "man's arts" and women seeking training in or performing these arts were seen to be "out of line," brazenly "bad." Geisha originated in this context, and they compounded this "badness" by not fulfilling mandatory social duties of wife and mother. Even today, many people associate the term "geisha" with "badness" whether or not they have had any experience with or knowledge about geisha, their history, or the traditional arts.

Confronting geisha "badness" requires us to unravel the unique world of the Japanese traditional arts with its historical views toward female performers, the complicated social hierarchies found within the *iemoto-ryū* system, the layers upon layers of interrelated genres of music and dance, and the expenses and training schedules that result. The understanding gained allows for the possibility of seeing the elements previously interpreted to be "bad" to be statements of independence and possibly even nonconformism. However, the Western "geisha-girl" image and its corresponding "badness" within Western societies will be no help to such inquiries whatsoever.

Notes

I thank the Midwest Japan Seminar, with the support of the Japan Foundation, for an opportunity to present this research at Loyola University Chicago.

1. However, any woman can be referred to (by herself or others) as "geisha," and there are women called "geisha" (such as "hot springs geisha" *onsen geisha*) that may not match the criteria I have just outlined. The definition provided here is what the traditional arts society holds, as conveyed to me during fieldwork interviews and interactions.
2. While the situation for modern Kabuki actors is very different (they do not have to pay to maintain their own theaters), actors prior to the twentieth century had to engage patronage in the same manner as do geisha.
3. Because patrons and geisha share common interests, intimate relationships sometimes develop. However, even if sex is involved, it would be incorrect to view these cases as sex-for-money arrangements.
4. Jodi Cobb, *Geisha* (New York: Alfred A. Knopf, 1995), 8–10.
5. Joan Pross Larson, *The Geisha Cookbook: Japanese Cookery for Americans* (Mount Vernon, NY: Peter Pauper Press, 1973), 5.
6. Arthur Golden, *Memoirs of a Geisha* (New York: Alfred A. Knopf, 1997).
7. "Pages Torn From the Memoirs of a Geisha. Mineko Distances Herself from Sayuri," (*The Japan Times*, March 12, 1999). Gion is a specific community in Kyoto.
8. Mineko Iwasaki with Rande Brown, *Geisha, a Life* (New York: Atria Books, 2002).
9. Frederick Joss, *Of Geishas and Gangsters: Notes, Sketches, and Snapshots from the Far East* (London: Odham's Press, 1962), 36.
10. Boye de Mente, *Some Prefer Geisha: The Lively Art of Mistress-Keeping in Japan*, (Rutland, VT: Charles E Tuttle, 1966), 39 and 97.
11. Iwabuchi Junko, "*Danna*" *to asobi to Nihon bunka* (Tokyo: PHP Kenkyūjo, 1996), 34.
12. Judith Briles, "Banish Geisha Nursing (and Other Female-Dominated Professions)," *The Briles Report on Women in Healthcare* (San Francisco: Jossey-Bass Publishers,

1994), 201; Katherine Roth, "As 'Geisha Chic' Hits U.S., One, 87, Scoffs," *The Japan Times*, June 22, 1999. *We are Ninja (Not Geisha)*, performed by Frank Chickens and produced by Kaz Records, 1984, "Geisha Boys and Temple Girls," *Pavement Side* (Performed by Heaven 17 and produced by Virgin Records, 1981).

13. Joyce Zonana, "The Sultan and the Slave: Feminist Orientalism and the Structure of Jane Eyre," in *Revising the Word and the World: Essays in Feminist Literary Criticism*, eds. VéVé A. Clark, Ruth-Ellen B. Joeres, and Madelon Sprengnether (Chicago: University of Chicago Press, 1993), 167–168.

14. Liza Crihfield Dalby, "Courtesan and Geisha: The Real Women of the Pleasure Quarter," in *The Women of the Pleasure Quarter*, ed. Elizabeth de Sabato Swinton (New York: Hudson Hills Press, 1996), 47–66.

15. From 1743 on, *odoriko* were sent to Yoshiwara just like unlicensed prostitutes. Mitamura Engyō, *Karyū fūzoku* in *Engyō Edo bunka* series, vol. 27 of *Engyō Edo bunko* series, ed. Asakura Haruhiko (Tokyo: Chūō Kōronsha, 1998), 186.

16. Ibid., 166.

17. Ibid., 218.

18. Ibid., 174.

19. Ibid., 174 and 185.

20. Ibid., 186.

21. A. Kimi Coaldrake, "Female *Tayū* in the *Gidayū* Narrative Tradition of Japan," in *Women and Music in Cross-Cultural Perspective*, ed. Ellen Koskoff (Urbana and Chicago: University of Illinois Press, 1989), 154.

22. Nakagawa Kiun, *Kyō warabe* (1658), found in Aketa Tetsuo, *Nihon hanamachi shi* (Tokyo: Yūzankaku Shuppan, 1990), 22.

23. Asai Ryōi, *Edo meishoki* (1662), found in Aketa, *Nihon hanamachi shi*, 22.

24. Benito Orlotani, *The Japanese Theatre: From Shamanistic Ritual to Contemporary Pluralism* (Princeton, NJ: Princeton University Press, 1995), 164–165.

25. Donald Shively, citing Hayashi Razan in "The Social Environment of Tokugawa Kabuki," in *Studies in Kabuki. Its Acting, Music, and Historical Context*, ed. James Brandon, William P. Malm, and Donald H. Shively (Honolulu, HI: University Press of Hawai'i, 1978).

26. Many students of the *ryū*, not just geisha, take the group name. However, the absence of a husband's name marks geisha as unique (most non-geisha female students eventually marry).

27. *Kikijōzu*, published during the An'ei period (1772–1780), as cited in Mitamura, 235.

28. Iwabuchi, *"Danna" to asobi to Nihon bunka*, 33–34.

29. Ibid., 44.

30. Ibid., 44–45.

31. Ibid., 215.

32. Mitamura, *Karyū Fūzoku*, 269–270.

33. Blind itinerant performers known as *goze* played *shamisen* and were female, however.

34. The *koto* is performed either as a chamber instrument (solo or in very small groups), or within the *gagaku* court ensemble.

35. However, some Kyoto geisha (particularly in Gion) also include the *koto* in their musical studies even today.

36. In 1983, an all-female Kabuki troupe (called *Musume Kabuki*) was founded in Nagoya in order to address this inequity. While not nearly as famous as the male Kabuki, it now carries the endorsement of the famous Nishikawa *ryū* (three of its members possess the Nishikawa name).

"An Illustration of Hibiya Park" by Yamamoto Shōkoku. Publication: *Fūzoku Gahō* No. 276, October 10, Meiji 36 (1903), Toyodo Branch Press.

3

Bad Girls from Good Families: The Degenerate Meiji Schoolgirl

Melanie Czarnecki

The sound of the bell was loud, and followed by the appearance of a beautiful, well-bred young lady of eighteen or nineteen. The long sleeves of her arrow-feather patterned kimono fluttering in the wind, her hair swept back at the sides and fastened with an innocent white ribbon, adorned in her *ebicha hakama*,[1] she was riding on a Dayton bicycle, her slender shoulders gliding.

—Kosugi Tengai, *Makaze koikaze*[2]

Introduction

Whether idly spending their free time in Hibiya Park,[3] experimenting with the latest hairstyles of the day, gossiping about the cute schoolboys they met on the street, or seriously poking their noses in schoolbooks, "bad" girls from good families did *not* embody the ideal of Japanese womanhood that their "good" girl classmates epitomized. Following the promulgation of the Girls' Higher School Order (*Kōtō jogakkō rei*) in 1899,[4] a small number of society's elite daughters found themselves newly situated within the liminal space of the girls' higher school. For the first time, wealthy daughters from the provinces were relocating to the city and intermingling with the natives at Tokyo's best higher schools. Many of these girls, away from the surveillance of family and servants for the first time, were able to form new communities comprised of both Tokyo residents and non-Tokyoites. Precisely due to its liminal characteristics, the fledgling system of the girl's higher school inadvertently provided the girls with opportunities extending beyond the walls of the classroom. Consequently, the emerging schoolgirl culture made it possible for the girls to position themselves in public spaces that had previously been off limits. As will become evident, it was their negotiation of these new spaces that largely defined their status as "good" or "bad."

The scope of this chapter is limited to a study of the small group of privileged schoolgirls studying at higher schools during the end of the Meiji period (1868–1912), primarily the years following the implementation of the Girls' Higher School Order. Therefore, any use of the term "schoolgirl/s" is specifically meant to denote those studying at higher schools at the end of Meiji, unless otherwise specified. At this time the schoolgirl was unilaterally represented by the daughters of Japan's most affluent families; those who could afford the exorbitant school tuition averaging 25 yen per year.[5] In 1903, the number of girls registered at Higher Schools had climbed to 25,719, up from the mere 3,020 recorded ten years earlier in 1893.[6] Considering, however, that the overall population of legally registered higher school aged Japanese female citizens was 2,116,058 in 1903[7] and that only 1.2 percent of those were registered at higher schools, it becomes obvious that lower-class girls who possessed neither the leisure nor funds to attend higher schools were eliminated from this schoolgirl coterie. Girls' higher schools were designed to function as a temporary space conducive to the conditioning of the ideal Japanese woman, who was expected to be a morally virtuous good-wife and wise-mother.[8] Accordingly, time spent at the higher school was intended to be a safe passage from childhood to adulthood preparing the girls for their future re/productive roles as the wives and mothers of Japan's soldiers (Sino-Japanese War 1894–1895, Russo-Japanese War 1904–1905). Their successful accomplishment of these roles ensured their "good" girl status. Sharon Nolte and Sally Hastings paint a detailed picture of who this "good" schoolgirl was envisioned to be. She was expected to dedicate herself to becoming a good-wife and wise-mother by pursuing "whatever employment and education would serve her family and the society." She was also to be filial, frugal, and possess feminine modesty.[9]

Conversely, this chapter attempts to illuminate how the "bad" schoolgirl was imagined. By the late-Meiji period the label "degenerate schoolgirl" (daraku jogakusei) had gained currency through Japanese reportage. Grounds for this pejorative branding were any and all actions of resistance against the state goal of churning out morally pure good-wives and wise-mothers. Any girl who strayed from this path risked becoming a model of degeneracy and the upper status level of the schoolgirl made even her slightest digression cause for scandal-mongering media to celebrate. We can imagine then the frenzied media reception of scandalous stories connecting schoolgirls to sexual deviance. The "bad" girl was depicted as a moral transgressor with no regard for the sanctity of feminine virtues, namely that of virgin purity. Her deviance may have been *seen* in her boldness to hold hands with schoolboys, or perhaps even steal a kiss. In extreme cases it was exposed by pregnancy or abortion. Or as one 1908 magazine article explains: Society's opinion of the schoolgirl as "shameless comes from her nonconformity to what has been considered the order of feminine beauty; chastity, elegance, and grace."[10]

Sensationalized images of degenerate schoolgirls in the media both reinforced and led to their stigmatization as a new bad-girl breed responsible for destabilizing the traditional foundations of patriarchal society. Damning images were ubiquitous; they came in the medium of graphics, plays, novels, songs, guidebooks, newspaper accounts, magazine articles, and conduct manuals. The media avidly depicted these sexual deviants as poster-girls for degeneracy. They signaled a warning to all good families of what dangers lurked in the environs of girls' higher education. Often schoolgirls accused of deviance would find their names in print, or as was the case with notorious schoolgirl Hiratsuka Raichō (1886–1971),[11] even their pictures. As such, the media was largely responsible for the construction of the degenerate schoolgirl. The intensity of such media interest raises a number of questions about the purposes these damaging images were intended to serve. We can also ask, in what ways were the late-Meiji schoolgirls imagined and through what vehicles was their so-called degeneracy diffused? Did their allegedly degenerate nature act as a necessary foil to the preservation of the good-wife–wise-mother? Did outright attacks condemning degenerate schoolgirls work to successfully conceal the underlying complexities enmeshed in their actions, that is, their own quests for intellectual and economic independence or their desire to consecrate meaningful unions guided by love, rather than arranged marriages?

While considering these questions I look at images of degenerate schoolgirls in a Meiji period graphic print, conduct manual, guidebook, novel, and newspaper in an attempt to uncover certain meanings both manifest and latent in the imagery associated with the schoolgirl and her so-called fall to degeneracy. We see how the circulation of visual representation through graphics often functioned to naturalize the ideologies of the day. It will become clear that conduct manuals and guidebooks served to construct and enforce appropriate deportment for girls, while novels and newspapers put manual and guidebook principles into narrative form, frequently providing the reader with peepshow thrills legitimatized in the guise of didactic tales of morality.

Good-Wife–Wise-Mother Ideology and the Socially Taboo

In July 1899, Education Minister Kabayama Sukenori (1837–1922) summarized the purpose of girls' higher education as the following: Girls' higher schools "exist for the nourishment of good wives and wise mothers. As a consequence, together with nourishing a warm and chaste character and the most beautiful and elevated temperament, it is necessary that they furnish the knowledge of arts and crafts necessary for middle to upper-class life."[12] Kabayama sums up the state's ideal image of the modern Japanese woman in a nutshell. She must be a good-wife–wise-mother aspiring middle/upper class virgin. Kabayama, however, neglects to mention the role they would serve in the larger scheme of things. As Rebecca Copeland informs us in *Lost Leaves: Women Writers of Meiji Japan*, "[T]he Woman Question (*fujin mondai*) became the focal point for reformers' efforts to 'civilize' Japan. 'Woman' was positioned as a metaphor for all that was backward and shameful in Japan. From an Orientalist regard, Woman signified Japan itself: a weak nation amidst superior Western powers."[13] Copeland goes on to say that educational "reformers believed that a nation's civilization could be measured by the status of its women" and that "[b]y improving the status of Woman, these men assumed, they could improve the status of the nation."[14] However, Japan's success in the Sino-Japanese War, predating the Girls' Higher School Order by four years, aided the transformation of its status from "weak" to "strong" on an international scale, while the status of Japanese women remained virtually unchanged. This exposed the mythical nature of Japanese reformers' earlier attempts to define its "backwardness" in terms of women's status.

The government's decision to finally support girls' higher education, some thirteen years after it had already acknowledged and implemented educational orders aimed at boys' higher education, suggests that along with the internalization of Western models for girls' higher education, the Japanese government was also stimulated by self-serving intentions. In a time when top-notch soldiers were in high demand and women were left to care for the home while spouses were out on the battlefield,[15] women's higher education, along with elementary and middle school education, served as an asset for the state.[16] Girls' higher schools were expected to foster a new sense of patriotism and to prepare the girls for their good-wife–wise-mother role. Girls were encouraged to reproduce many baby boys, raising them to become strong and powerful soldiers in Japan's military. It is within this historical context that girls' higher education was formalized. In other words, girls' education was instituted overwhelmingly for the sake of a prosperous and powerful nation (*kokka fukyō*). While it is imperative to emphasize the new Meiji view of motherhood that was espoused through good-wife–wise-mother ideology, it is just as important to underscore what Kabayama meant when he claimed that girls' higher education was meant to nourish a "chaste character."

The fact that educational curriculums at many girls' higher schools incorporated teachings on morality was not due exclusively to the influence of Confucianism or the newer teachings of Christianity. The Meiji government had its own reasons for expecting women to serve as an emblem of moral propriety by remaining chaste until marriage. Under the 1871 family registration law, the patriarch was the household head and controlled all family property which the eldest son was expected to inherit. Newly instituted laws made adultery a punishable crime in 1908, suggesting government measures to ensure that all prospective heirs were indeed biological offspring. While son-less families often adopted an heir, the importance placed on the birth of a blood-related male for inheritance purposes became a top priority. Therefore, the modern belief that blood from previous sexual partners remained in a woman's body after intercourse, thus affecting the paternity of future offspring, resulted in a new demand for virgins.

As Kawamura Kunimitsu notes in *The Modernization of Sexuality* (Sekushuariti no kindai), this conviction called for "women to remain virgins until marriage and protect their chastity in order to maintain that they had 'pure blood' and there was a 'pure bloodline,' free from the 'mixed blood' that occurs from sex with multiple partners."[17] This belief paved the way for the Western doctrine of virgin purity and spiritual love to become a prerequisite for a truly modern relationship in the civilized and enlightened world of Meiji Japan. Prior to the Meiji period, Edo-ites (1600–1868), with the exception of those belonging to the samurai class, were able to engage in multiple sex relations freely. By the late 1880s, however, the Western doctrine of virgin purity had spread through the writings of influential critic and poet Kitamura Tōkoku (1868–1894) to be embraced by a large audience of Japanese women, including the schoolgirl.[18]

Innocent Good-Girls in Print

The schoolgirl was initially seen as little more than a candy-coated symbol of modernity. The fact that these good girls had been largely hidden away from the public eye before passing through the gates of the higher school, worked to enhance their movie-star like mystique. These well-bred young ladies were meant to impress the West. They were an emblem of Japan's progress made manifest in a tennis-playing, English-speaking, educated beauty. A look at the 1903 print "An Illustration of Hibiya Park" (*Hibiya kōen no zu*) appearing in the popular magazine *Fūzoku gahō* (customs and manners pictorial), and reprinted at the outset of this chapter, may help to envision more clearly the decorative element the schoolgirl lent to the fabric of society. This print is the first to appear in Honda Masuko's account of the Meiji schoolgirl, *A Chronology of the Schoolgirl: The Meiji Period in Color* (Jogakusei no keifū: Saishoku sareru Meiji). Honda's title of the print however differs from that used in *Fūzoku gahō*. She instead goes by the title "Hibiya kōen ni ikō hitobito" (People resting in Hibiya Park).[19] As we can see, the print embellishes the modern: electric lights, a baby carriage, a balloon, a bicycle, Western garb and more. On the right side of the print we have two fashionable Western women, to the left of which stands a kimono-clad Japanese woman clutching a baby. She in turn is juxtaposed by the divas of the scene: two chit-chatting, *hakama*-sporting schoolgirls. The depiction of the two schoolgirls appears to illuminate their elite, yet inconsequential status. This can be said especially when contrasted to the other discernable Japanese women in the print who seem to be connected by a common bond. That shared bond is their gendered social role as childcare providers. Indeed, a closer look confirms that while the schoolgirls' hands are engaged with *books* (inside a wrapping cloth), the other women's hands are occupied with *children*. They are busy pushing baby carriages, comforting crying babies, and clenching the smaller hands of wayward children. Contrary to what Honda's title for the print would suggest, these women are far from *resting*, they are engaging in manual labor. The only people to be found truly resting are the Western women, men, and the schoolgirls, who Honda suggests were *seen* as nothing more than " 'privileged outsiders' unconnected with either life or living, ethereally sweet and innocently passing the time."[20]

Yet, while the schoolgirls may have been dubbed "prestigious outsiders," it does not appear that they were initially seen as a threat to society. Instead, as Honda suggests, they were viewed as "ethereally sweet and innocent." Sure enough, when paging through other issues of 1903 and 1904 copies of *Fūzoku gahō*, we find other images of schoolgirls to be equally innocuous, with the exception of one print depicting two schoolgirls riding on bicycles in the January 1, 1904 issue, which suggested a flair for the wilder side of schoolgirl culture. These representations clearly construct the schoolgirl as the Japanese female Other, the "privileged outsider." She escapes the struggle of commerce, farming, natural disasters and war, or working as a geisha that women in the other prints characterize. Instead, *Fūzoku gahō* offered prints of schoolgirls joyfully taking part in school sports and art festivals, or dreamily viewing autumn foliage in surreal scenes that make it easy to forget Japan was, at that time, at war with Russia. These depictions served to naturalize the state ideology that privileged the existence of schoolgirls to those of her *lesser* counterparts. At the same time, they reinforced the idealized image of Japan's modern sweetheart that the state hoped would appeal to the West.

Behind the images adorning these graphic prints however, a demonic version of this so-called innocent schoolgirl was beginning to take shape. Before long, the "ethereally sweet and innocent" schoolgirl was transformed into a corrupter of public morals. A newspaper depiction of schoolgirls in Hibiya Park would likely cast a very different impression than the one described above. No longer would innocent schoolgirls, with books in hand, radiate in the daylight, but rather they would transform into degenerate girls, shadowed by the moonlight, their books replaced by the hands of their schoolboy lovers. In fact, Hibiya Park eventually became a public space notorious for sexual dalliances among Tokyo's schoolgirls and boys.

Demon Winds Love Winds: The Degenerate Schoolgirl in Fiction

Increasing opportunities for schoolgirls and boys to meet brought a new dimension to female/male relations in late-Meiji. Romances between schoolgirls and boys captivated both society's curiosity and scorn, as not all relationships were able to maintain their promise of chastity. Newspaper stories of good-girls-gone-bad spurred the development of a new literary genre: degenerate schoolgirl discourse and the popular writers of the day were there to capitalize on the public's fascination with these elite lovebirds. Kosugi Tengai (1865–1952) was one such writer. His novel, *Makaze koikaze* (Demon Winds Love Winds), was serialized in the *Yomiuri Newspaper* from February 25 to September 16, 1903. *Demon Winds Love Winds* enjoyed widespread popularity, turning into one of the best sellers of the Meiji era. One of the main attractions of this novel was Kosugi's decision to cast the late-Meiji schoolgirl as its heroine. As early as 1887, schoolgirls were the topic of literary imagination and had come to replace the geisha and prostitute as the literary heroine of choice. The public emergence of the schoolgirl intrigued writers who were eager to bring this new heroine to life on paper. The scandal-laden love relationships between students depicted in the novels had the added advantage of luring in the readers.

From Meiji novels that introduced schoolgirl heroines, readers were able to glimpse the liminal world where schoolgirls challenged so-called normative social structures. One construction of the degenerate schoolgirl brought to life by Kosugi was nineteen-year-old Hagiwara Hatsuno, the lead character in *Demon Winds Love Winds*. (Her debut scene is quoted at the beginning of this chapter.) Hatsuno's tale provides one account of the type of degenerate schoolgirl who flirted with sexual deviance. At the same time, her ambitious drive to graduate from the elite Imperial Girls' College (*Teikoku Joshi Gakuin*)[21] is fueled not by her desire to possess the credentials of a modern wife, but instead by her yearning to secure a job that will lead to economic independence. Her disregard for the higher school's objective to mold good-wives and wise-mothers doubly casts her in the role of degenerate schoolgirl. Hatsuno is from a well-to-do family, but her family tree is marred by the fact that she and her younger sister, Nami, are the daughters of her father's mistress. Her life is complicated when her elder half brother, Kichibei (the son of her father's official wife), rises to the position of head of the family after her father's death and makes no effort to hide his annoyance at having a schoolgirl for a sister.

Almost immediately after the story opens, Hatsuno finds herself in an economic bind after she is involved in a bicycle accident that lands her in the hospital.[22] As her brother refuses to pay the hospital bill, she is obliged to leave the hospital against her doctor's orders and unsuccessfully attempts to sell her damaged bicycle to cover the cost. The proprietress of the boardinghouse where Hatsuno resides, Oshufu, arranges for an acquaintance, a young man and artist named Tonoi Kyōichi, to lend Hatsuno the money. Hatsuno reluctantly accepts it. When she goes to the hospital to settle the account, however, Hatsuno finds that the bill has already been mysteriously paid and promptly returns the borrowed money to Tonoi. She later learns that Togo Natsumoto, the fiancé of her classmate Yoshie Natsumoto, had covered her debt.

Calamity occurs when a confrontation based on a misunderstanding between Hatsuno and her brother results in Hatsuno losing his much needed financial support. After her brother claims that the family fortune is more important to him than his two half-sisters, an irate Hatsuno swears:

> From now I will not rely on you for anything, not even my school tuition. Nami and I would rather die than be at your mercy! To which Kichibei replies, "Well, you've said it now. You better not forget it." And Hatsuno answers, "I will not forget. After all, I am an educated woman."[23]

Unquestionably proud of her schoolgirl status, Hatsuno is nonetheless left in a state of financial duress. Her wealthy classmate Yoshie, offers economic assistance, but their sister-like relationship is later compromised when Yoshie's mother, Baroness Natsumoto, walks into a room which her husband, the Baron, has just exited, to find Hatsuno quickly retying her kimono sash (after an unwelcome assault by the Baron). Demanding an immediate conference with Hatsuno, Baroness Natsumoto inquires:

> Did you come here to perform this kind of lascivious act? I hear that schoolgirls nowadays who are strapped for cash to pay tuition will often prostitute themselves. It appears that you are one of those, does it not?[24]

Distraught over the Baron's attack and the Baroness' offensive accusations, Hatsuno becomes firmer in her belief that the only way she can be financially independent is if she graduates from College with qualifications for a good job. Hatusno, who is highly regarded for her beauty and genius, surely could have found a new benefactor. In fact, Oshufu, went to great lengths to convince Hatsuno to accept the artist Tonoi's numerous proposals, but Hatsuno refused to sacrifice her dignity in a loveless union for the sake of economic protection. Here we glimpse a crucial aspect of the new and modern schoolgirl mentality—a devotion to the sacredness of spiritual relationships. Not only did Hatsuno assertively take her future into her own hands, proud of the potential earning power of her education, but she also upheld her belief of not entering into a relationship without love. As a result, she battled through desperate times, but in doing so, was able to preserve her integrity. And so, with only a few months left until her final exams, Hatsuno throws herself into her studies. By this time, however, an exhausted and ill Hatsuno is diagnosed with beriberi. The plot then thickens as Togo, whom the son-less Natsumotos have adopted as their heir, admits that he loves Hatsuno more than his intended. "I love that girl. I love her more than Yoshie,"[25] he declares to his biological mother. Afterwards, rumors begin to spread that Togo has been disowned by the Natsumotos and that although Yoshie has been informed of his relations with another woman, she is not told immediately that the other woman is her beloved Hatsuno. The climax of the melodrama occurs as a crying Togo confesses his love to Hatsuno.

> I never told you this, but it wasn't just yesterday or today that I fell in love with you. It was many years ago . . . It was probably the first time that I ever saw you.[26]

This dramatic love scene, risqué by 1903 standards, culminates with Togo placing his lips on Hatsuno's and asking her to be his wife. We learn through the narrator that this is the first time in Hatsuno's nineteen years that she comes to know the lips of a man. "Her body quivered, her heart was thumping, believing that she would lose her breath, she happily indulged in the moment, feeling as though she was flying in a warm dream world."[27] Although distraught over betraying Yoshie, Hatsuno nonetheless chooses her love for Togo over that for her friend and agrees to marry him. According to Saeki Junko, the act of the "kiss" was viewed as an expression of love not belonging in the same category as "physical relations," but rather adapted from the Western tradition as a type of greeting or parting.[28] From this perspective, the spiritual purity of Togo and Hatsuno's relationship continues to be safeguarded despite their union of lips. Still, it is likely that some readers of the day imagined a more than purely spiritual relationship between the two, as the "kiss" had formerly been connected exclusively to sexual relations in Japan.

This love triangle continues to haunt everyone involved and Yoshie becomes sick with depression at the knowledge of Togo having another love interest. Togo, who owes his entire privileged upbringing to the Natsumoto family eventually succumbs to pressures from both his biological and adopted family to break his marriage proposal to Hatsuno and return to Yoshie. Unbeknownst to Togo, Hatsuno, while eavesdropping, hears him plead with his adoptive family, "Please, forgive me for my wrongdoings. I want nothing more than for you to allow me to marry Yoshie."[29] In her powerful soliloquy that follows Hatsuno declares:

> I was not born for the convenience of others, to have my feelings toyed with, to be abandoned because of circumstance. And still, to have to protect my honor and act

docile, what is that all about? Such an obligation does not exist, it does not. And if it did, who could endure it? I am not a machine! I am not a slave! I am an individual human being! I am an educated woman![30]

Here we discover Hatsuno's dismay over Togo's weakness in the face of adversity, and once again, the sense of pride that a growing number of educated schoolgirls by the end of Meiji were starting to display and society consequently fear. Hatsuno herself was unwilling to bow to circumstance and she is angered to find that Togo does not possess the same strength. Nonetheless, Hatsuno, in a state of disbelief, cannot stop longing for Togo to appear and assure her that what she overheard was only an act. While castigating herself for falling in love with Togo, she remains unable to completely abandon the hope that their love will be realized in a marriage union. The narrator gives us access to Hatsuno's internal thoughts as she grapples with both the thought of her impending examination and relationship with Togo. "Even though this exam is of utmost importance, I can't bear to give up this love, not now. I mean, I can't possibly do well on this exam if I'm broken hearted . . . Something has to happen, and after it does, then I will study, it won't be too late."[31] Here, Hatsuno carries on an internal struggle between her heart's desires and her responsibility as a student, with love winning over pragmatism. Not only does she give in to illicit romance, but she turns to betraying her best friend, too.

In hopes of pushing Yoshie out of the picture, Hatsuno deceives her into going on a wild-goose chase to search for Togo in a faraway town, although he is really right there in Tokyo. Her conscience however gets the best of her and she hurries to the port where Yoshie is waiting to board a ship and confesses her crime. Begging Yoshie's forgiveness Hatsuno collapses on the ground. Obviously ill, she is taken to a hospital. On her deathbed, Hatsuno takes her last breath while holding the hands of both Yoshie and Togo. Suffering from heart failure brought on by beriberi, the beautiful and brainy heroine finally succumbs. Kosugi closes the novel with the romantic image of a wedding ceremony for Yoshie and Togo and his message is clear. Hatsuno's degeneracy, which culminated in her death, served as a foil to the preservation of morality and "goodness," as was depicted in large through the representation of Yoshie, a refined, wealthy young lady, who is eager to consummate her wedding vows and become the good-wife and wise-mother her family and state expects. Her only "fault" appears to be her occasional tendency to disobey her mother in circumstances concerning Hatsuno, an act of filial defiance of which Hatsuno is seen as the perpetrator, once again underscoring her degenerate nature.

Boarding House Bad Girls

Many schoolgirls coming from outside of Tokyo sought lodging at boardinghouses. Before long boardinghouses became synonymous with scandal, as condemning images of degenerate schoolgirl boarders began to spread. This phenomenon is reflected in the following passage from *Demon Winds Love Winds*.

Up until last spring, before schoolgirl scandal became the talk of the town, more than ten rooms were filled in this boardinghouse accredited by the Imperial Girls' College. Good business continued, but then when school regulations changed and accreditation was revoked, it became lonely like it is now. Except for two or three girls with permission to stay on, there are no boarders from the Imperial Girls' College here, only some girls from no-name schools, nurses, and midwives.[32]

Hatsuno, originally from Chiba, initially resides here. Following her bicycle accident, mentioned in the previous section, the boardinghouse mistress, Oshufu, takes Tonoi (who has not yet met Hatsuno) into the privacy of her room to show him the bicycle. The moment Tonoi enters the private space of Hatsuno's room he becomes a trespasser. Emphasizing its lack of feminine décor he exclaims that the room looks "exactly like a schoolboy's room."[33] Then, upon hearing that Hatsuno is behind with her rent, he remarks that she "surely must be one of those degenerate schoolgirls" and wonders whether she "started man-chasing."[34] Yet it is such thoughts of Hatsuno's degeneracy that captivated Tonoi, exciting him to attempt to uncover clues about this mystery girl. In the process, he re-constructs her identity based on what he uncovers, on what he *thinks* he sees. Not only is he a trespasser, but he is also a Peeping Tom of sorts. As he digs through her drawers, he pulls out a small bottle of pills, inquiring "What are these? Probably syphilis medicine."[35] Next he finds pawnshop receipts and his suspicions increase. The image of Hatsuno that Tonoi creates for the reader represents one example of the damaging types of stereotypes that were circulating at the time. Kosugi's characterization of Hatsuno is said to have been modeled on a real life example of a schoolgirl who overdosed on morphine.[36] Such real life examples often provided the stimulus for fictional accounts of the late-Meiji schoolgirl, thus acting to naturalize the notion that all schoolgirls had the latent potential to become degenerate. Like Tonoi, the public was eager to speculate on the "badness" of schoolgirls and embellish wayward stories with little evidence. As it turns out, Hatsuno's "syphilis medication" was actually sleeping pills.

Surmounting scandal connecting schoolgirl degeneracy to boardinghouses led schools to prohibit students from taking up residency in these so-called houses of vice. Campaigns ensued and educators took pains to enlighten the public on the roots of boardinghouse evil. While many people held the schoolgirl responsible for her own actions, others tended to sympathize with her predicament, blaming the seductive prowess of schoolboys for the degeneracy of the naive victims and placing heavy responsibility on the fathers' obligation to protect their daughters from moral corruption. As Kawamura Kunimitsu points out in *A Maiden's Body* (Otome no shintai), "under the Meiji Civil Code, unwed women (daughters) and wives were entities to be protected by the household patriarch. Throughout a woman's life, her 'virginity' and 'chastity' were to be regulated—overseen through her relationship to the patriarch—the father or husband . . . The actual owner and manager of 'virginity' and 'chastity' was not the woman herself, but the father or husband."[37]

Some educators provided conduct manuals on how to stay on the straight and narrow. Sumire Hoshino's *Modern Schoolgirl Thesaurus* (Gendai jogakusei hōkan), published in 1906, offered schoolgirls advice on everything from how to put on make-up, devise the best study methods, choose friends, and of course, avoid degeneracy. She includes a section dedicated to boardinghouse dangers, condemning them as breeding grounds for moral corruption and warns, "while boardinghouse living is very convenient on the one hand, dreadful sins are lurking."[38] Similarly, Otsuki Hisako's *A Guide for Schoolgirls Studying in Tokyo* (Tokyo joshi yūgaku annai), published in 1907, cautions:

> Those living in rented rooms will find out soon enough that they provide the basis for degeneracy. And those living in boardinghouses will never be able to remove the stains that will ultimately darken their entire lives. Heed my warning.[39]

The prevalence of damaging images of boardinghouses, followed by their condemnation as unfit to house schoolgirls, caused the displacement of many girls who had bonded with other schoolgirl boarders from across Japan, enriching their cultural experiences. As a result, the schoolgirls' occupation of this liminal space was short-lived. While the possibility that boardinghouses could facilitate physical relations between schoolgirls and boys was a valid concern at a time when a daughter's marriage prospects hinged on her virginity, the issue of same-sex physical relations did not garner the concern that potential heterosexual relationships did. As Sarah Frederick demonstrates in chapter 4 of this volume, lesbian love within the confines of boardinghouses did indeed occur.

Newspaper Depictions of Degenerate Schoolgirls

Masaoka Geiyō's *The Other Side of Newspaper Companies* (Shimbunsha no rimen), published in 1901, details how the publication of scandalous material was a tactic used to reel in the readers. "In order to capture the attention of reading audiences," Masaoka writes, Japanese journalists had to "take a course that would expose people's abject circumstances and immoral behavior. Daily newspapers are actually made with this plan in mind. Expose the ignoble gentlemen! Exaggerate your writing and make it beguiling! These are the requests made by editors day upon day."[40] According to Masaoka, this was the strategy employed by most of the smaller tabloids. He continues to tell us that unscrupulous journalists "take no end of pleasure when it comes to the immorality of the wife of a powerful family, a well-bred young lady, or a widow."[41] Surely, this means of detraction was not limited only to the smaller tabloids. Even the larger, more accredited newspapers won huge popularity among their readers by offering serialized stories of the morally questionable, as was the case with *Demon Winds Love Winds*, among others.

Writers accused of corrupting public morals or writing about carnal desire solely for its shock value were subject to censorship. As Nakayama Akihiko suggests, "the problems surrounding the relationship between 'naturalism' and 'carnal desire' that began to emerge on the social scene in the late-Meiji period caught the eyes of authorities and faultfinding writers who worked to purge society of indecent literature."[42] However, he informs us that with regard to the criteria for banning material, there was no formalized structure intact to discern whether certain literary content was acceptable or not in late-Meiji. At the same time he does suggest that there was a tendency for novels composed in a nebulous style to face censure primarily due to assumptions about their power to stimulate the imagination of young boys and girls which in turn might prompt degenerate behavior.[43]

Works that can be categorized as degenerate schoolgirl novels such as *Demon Winds Love Winds*, Tayama Katai's (1871–1930) *Futon* (The Quilt, 1907) and Oguri Fūyō's (1875–1926) *Seishun* (Youth, 1906) contained passages that would be read as explicitly immoral and sexually graphic.[44] They were likely able to escape censure due to the moral lessons imbedded in their texts. However, as *Demon Winds Love Winds* was eventually removed from the National Diet Library in 1907, we are led to believe that during its release in 1903, government attention was focused on the approaching Russo-Japanese War rather than the policing of novels. In these novels, all three degenerate schoolgirl heroines meet a tragic end. *Demon Winds Love Winds'* Hatsuno, of course, succumbs to beriberi. *The Quilt's* Yoshiko is forced to leave Tokyo and return to her parent's home in the country-side after it is discovered that she had a physical relationship with her schoolboy lover. *Youth's* Shigeru's sexual relationship with her schoolboy lover leads to abortion and disgrace. The allegorical nature of these stories was perhaps seen as an effective warning to would-be sexual transgressors, sending them the message "play and you will pay." The device worked to present readers with the vicarious thrill of transgression in ways that did not question prevailing morality.

The *Yokohama Daily Newspaper*, perhaps in an attempt to capitalize on the popularity of serialized schoolgirl stories, printed *Tales of Degenerate Schoolgirls* (Daraku jogakusei monongatari) in 35 installments from September to November 1905. Appearing to be a nonfiction account of late-Meiji schoolgirl life, this series not only served as a cautionary marker to those "good families" whose daughters were in potential danger of falling from the grace of society, but also provided tantalizing tidbits that surely lured in a large reading audience fond of gossip. At the outset, the series' anonymous writer, Mr/s. X, who claims to be an educator, immediately lunges into an explanation of the connection between schoolgirl degeneracy and boardinghouse life, arguing that schoolgirls living in school dormitories or under the strict surveillance of their fathers or brothers are the least likely to be reproached for bad behavior. This claim reinforces the previously mentioned belief that it was the patriarch's responsibility to protect his daughter's virginity. Taking us for a peek inside the boarders' rooms, X describes the following scene.

> Their kimono and *hakama* lay scattered about carelessly. Their rooms look just like one of those used clothing shops that you find on some big street. In the morning, they leave without cleaning up after themselves, their dirty trays and bowls left about. Even boys aren't this bad. Of course, I am not saying that all schoolgirls follow this pattern, but there are a great many who apply. They come back from school and sprawl themselves out in the middle of their room, carrying on about mindless matters, speaking nonchalantly about that which is offensive to the ear and bragging about the number of schoolboy acquaintances that they have. Naturally, such behavior inevitably leads to immorality.[45]

Providing readers with hard-core stories of real-life degeneracy, X mentions that s/he was informed by one boardinghouse proprietress that when the girls go out for the evening they often do not return until the next morning. We learn, as well, that some girls from good families who were raised as "butterflies" and "flowers" end up as maidservants or mistresses.[46] For example, "degenerate schoolgirl Matsumoto became pregnant with her schoolboy lover's child and was forced to quit school after covering the expenses of the birth left her penniless. Wanting money to go back to school she chose to become some man's mistress."[47] These attacks, while emphasizing the schoolgirls' degenerate deed, ignore the underlying issue at hand. Namely, that the desire for an education was such that some girls would willingly prostitute themselves to obtain one. This recalls the Baroness Natsumoto's icy accusation toward Hatsuno, "I hear that schoolgirls nowadays who are strapped for cash to pay tuition will often prostitute themselves." Rather than criticize the state for not taking measures to provide needy students with provisions, society instead placed the blame on the girls and the lengths they would go for an education.

Conclusion

The target audience of these cautionary accounts of schoolgirl degeneracy ranged from the schoolgirls themselves to their families, especially their fathers. Still, when considering that the transmission of newspaper and magazine articles to families in the provinces was unlikely, we are led to believe that fans of erotica were also targets. Degenerate images of schoolgirls were expected to serve as cautionary markers while they simultaneously provided racy subject material. The transformation of the schoolgirls' initial image as an "ethereally sweet and innocent" symbol of modernity to a symbol of "moral degeneracy" can be traced to the visible onset of their deviation from patriarchal norms, manifest in the platonic-turned-sexual relationships they had with schoolboys, as well as their attempts to carry out social roles beyond the confines of the domestic sphere. Indeed, their attempt to transgress the confines of the domestic sphere was interpreted as an attack on the state-proclaimed dogma of female essentialism that glorified the innateness of motherhood and was credited with upholding a "prosperous and powerful" state. Still, as state ideology binding girls' higher school teachings to motherhood essentialism was built on a mythical platform, it was bound to break down. Hence the subversive potential of transgressing schoolgirls was gradually realized because the liminality of girls' higher school culture made it possible for the girls to cultivate a growing consciousness that led many to question the orthodoxy of state ideologies in matters that concerned them. Taking charge of their own lives while negotiating public spaces that had previously been inaccessible served to dismantle the myth that their only place was in the home. No longer willing to passively accept their socially constructed roles like "good" girls, many assertively engaged their newfound freedom by endeavoring to create new meanings for their "self-establishment and public advancement" (*risshin shussei*),[48] a risk they were willing to take, even if it meant becoming an outcast in the "bad" girl camp.

Notes

I thank Akihiko Nakayama for his assistance in helping me to locate materials and patience in answering my many questions. I also thank the Bad Girl editors and contributors of this volume for their careful readings and constructive suggestions.

1. A long, pleated, maroon-colored divided skirt worn over the kimono.
2. Kosugi Tengai, *Makaze koikaze* vol. 1 (Demon winds love winds, vol. 1) (Tokyo: Iwanami Shoten, 1999), 9.
3. Modern park opened in March 1903.
4. See Ministry of Education, Science and Culture, *Japan's Modern Educational System: A History of the First Hundred Years* (Tokyo: Printing Bureau, Ministry of Finance, 1980), 118–121.
5. Average tuition by 1907 standards. See Hoshino Sumire, *Gendai jogakusei hōkan* (Tokyo: Nihon Tosho Sentā, 1992), 197–206. In 1907, the starting monthly salary was 50 yen for a government official and about 12 yen for a policeman. See Dōmei Tsūshinsha, *Nihon no bukka to fūzoku 130 nen no utsurikawari* (Tokyo: Dōmei Tsūshinsha, 1997), 564–593.
6. Statistics Bureau, Management and Coordination Agency, *Jinkō tōkei sōran* (Tokyo: Tōyō Keizai Shinpōsha, 1985), 80.
7. Ministry of Education, *Gakusei 80 nenshi* (Tokyo: Ōkurashō, 1954), 1049.
8. For various interpretations of good-wife–wise-mother ideology, see Koyama Shizuko, *Ryōsai kenbo to iu kihan* (Tokyo: Keisō Shobō, 1991).
9. Sharon Nolte and Sally Ann Hastings, "The Meiji State's Policy Toward Women, 1890–1910," in *Recreating Japanese Women, 1600–1945*, ed. Gail Lee Bernstein (Berkeley: University of California Press, 1991), 171.
10. "Jogakuseikan." *Shumi* (March 1908): 17–23.
11. While studying English at Seibi Girls' Higher School in 1908, Raichō was publicly condemned for her so-called love-suicide attempt with Morita Sōhei. See Hiroko Tomida, *Hiratsuka Raichō and Early Japanese Feminism* (Leiden-Boston: Brill, 2004), 115–137.
12. Ministry of Education, *Japan's Modern Educational System*, 119.
13. Rebecca L. Copeland, *Lost Leaves: Women Writers of Meiji Japan* (Honolulu: University of Hawai'i Press, 2000), 11.
14. Ibid.
15. The Meiji government's revision of the Conscription Ordinance in 1883 made exemption difficult and under the Meiji Constitution of 1889 universal military conscription was established and all subjects were amenable to service. See Shinotsuka Eiko, *Josei to kazoku: Kindaika no jitsuzō* (Tokyo: Yomiuri Shimbunsha, 1995), 49–51.
16. In 1903 2,251,051 girls were registered at elementary schools and 1,288 at middle schools. See Ministry of Education, *Gakusei 80 nenshi*, 1044–1047.
17. Kawamura Kunimitsu, *Sekushuariti no kindai* (Tokyo: Kōdansha, 1996), 123.
18. For a discussion on Kitamura Tōkoku, see Janet Walker, *The Japanese Novel of the Meiji Period and the Ideal of Individualism* (Princeton: Princeton University Press, 1979).
19. Honda Masuko, *Jogakusei no keifū: Saishoku sareru Meiji* (Tokyo: Seidosha, 1990), 11.
20. Ibid., 10.
21. Modeled on Japan Women's University (1901).
22. American models cost as much as 250 yen.
23. Kosugi, *Makaze koikaze* vol. 1, 163.
24. Ibid., 205.
25. Kosugi, *Makaze koikaze* vol. 2, 163.

26. Ibid., 193.
27. Ibid., 194.
28. See Saeki Junko, *"Shiki" to "ai" no hikaku bunka shi* (Tokyo: Iwanami Shoten, 1998), 156–167.
29. Kosugi, *Makaze koikaze* vol. 1, 264.
30. Ibid., 268.
31. Ibid., 284–285.
32. Kosugi, *Makaze koikaze* vol. 1, 23.
33. Ibid., 36.
34. Ibid., 38.
35. Ibid., 41.
36. Kosugi Tengai, *Meiji Taishō bungaku zenshū dai 16 maki* (Tokyo: Shunyōdō, 1930), 633.
37. Kawamura Kunimitsu, *Otome no shintai: Onna no kindai to sekushuariti* (Tokyo: Kinōkuniya Shoten, 1994), 223.
38. Hoshino Sumire, *Gendai jogakusei hōkan*, vol. 3 in *Kindai Nihon seinenki kyōkasho sōsho* (Tokyo: Nihon Tosho Sentā, 1992), 13.
39. Otsuki Hisako, *Shinsen Tokyo joshi yūgaku annai*, vol. 9 in *Kindai Nihon seinenki kyōkasho sōsho* (Tokyo: Nihon Tosho Sentā, 1992), 4.
40. Masaoka Geiyō, *Shimbunsha no rimen* (Tokyo: Shinseisha, 1901), 122.
41. Ibid., 125–126.
42. Nakayama Akihiko, "Shōsetsu 'Tokai' saiban no gingakei. Kūhaku no seijigaku," in Mitani Kuniaki, ed. *Kindai shōsetsu no "katari" to "gensetsu"* (Tokyo: Yūseidō, 1996), 55–81.
43. Ibid.
44. Tayama Katai, *Futon* (Tokyo: Iwanami Shoten, 1998), Oguri Fūyō, *Seishun* (Tokyo: Iwanami Shoten, 1994).
45. "Jogakusei daraku monogatari." *Yokohama Mainichi*, September 20, 1905.
46. Ibid., October 16, 1905 and November 17, 1905.
47. Ibid., November 17, 1905.
48. Rebecca Copeland explains that for women to be candidates for *risshin shussei* they were required to be properly educated and properly enlightened to their own unique role in the family, the nation, and the world. See *Lost Leaves*, 15. This role, of course, was the embodiment of good-wife–wise-mother ideology.

Dust cover image by Nakahara Jun'ichi, originally published in the 1939 edition of *Hana monogatari* (*Flower stories*); re-published in Kokushokan edition in 1995. With kind permission of Sohji Nakahara and Himawariya.

4

Not That Innocent: Yoshiya Nobuko's Good Girls

Sarah Frederick

We might expect that a writer who was openly lesbian in 1920s Japan would automatically qualify as a "bad girl." Yoshiya Nobuko (1896–1973) in many senses questions such an assumption. She prolifically created novels for serialization in newspapers and women's magazines from the 1910s to the 1970s, and throughout most of her writing life she lived with a female partner Monma Chiyo, and this became widely known. It would seem likely that her fiction would defy the gender norms of her time and extol bad girls of various sorts. Yet the female characters of her girls' fiction and romance novels exhibit hyper-typical images of "good girl" femininity rather than subvert them, and critics frequently describe her writing style as flowery (*bibunchō*). On the surface at least, her descriptions of pure and clean girls and women hardly overturn stereotypes and ideals of girlishness or feminine virtue. However, Yoshiya's celebration of ultra-feminine virtue and emotional intensity did work as a criticism of and resistance against gender norms and the family system during her lifetime, and her writings help to complicate our sense of what resistance or badness might be. In particular, Yoshiya used the flexibility of fiction to construct alternative gender expectations in a way that could appeal to a broad range of readers of various sexualities and genders.

Yoshiya Nobuko

Yoshiya Nobuko has rarely been translated into English and has only recently become the subject of literary criticism in Japan or elsewhere.[1] This is primarily because of her status as a popular writer, which placed her fiction outside the circle of critically acclaimed writers, while ignoring her influence on an entire generation of readers. She was one of the best-selling writers in twentieth century Japan and has been well known in Japan since the late 1920s. She had connections to many writers better known outside Japan, including people such as Tokuda Shūsei (1871–1943), Okamoto Kanoko (1889–1939), Hayashi Fumiko (1903–1951), and various members of the Bluestockings.[2] Her writings also provided material for adaptation to film, radio, Takarazuka theater, television, and manga, and consequently have had a reverberating effect on Japanese popular culture. A recent rather bulky biography of Yoshiya by contemporary woman writer Tanabe Seiko has created a resurgence of interest that has helped to trigger both critical attention and a re-release of many of her major works, including a series aimed at girls, *Flower Stories* (Hana monogatari), *Two Virgins in the Attic* (Yaneura no ni shojo), *Mrs. Ban* (Ban sensei), and *Forget-Me-Not* (Wasurenagusa), many illustrated by Nakahara Jun'ichi whose style influenced girls' comics.[3]

Yoshiya remains interesting to both scholars and the public for a number of reasons. Of course, the sheer number of her readers makes her sociologically important. The magazines in which she serialized during the late 1920s published close to one million copies, and her popular novels in postwar Japan were usually among the best sellers, with *The People of the Ataka Family* (Atakake no hitobito) selling over five million copies in the early 1950s. In a 1946 survey by Mainichi Shimbunsha titled, "Who is your favorite writer?" Yoshiya was the highest ranked woman writer, coming in fifth right after Natsume Sōseki.[4] It is hard to find a woman who read fiction before the 1970s who has not read one of her works. Tanabe recounts how the current empress of Japan, when asked what she had read in the early postwar, lit up with glee and listed off several Yoshiya novels.[5] Her fiction then tells us a great deal about what girls and women read in the last century, and perhaps something about what appealed to them. The rich descriptions and contemporary settings of most of her works mean that they engage in obvious ways with the eras in which they were written.

The emotionally rich and descriptive style of her writing influenced a number of important Japanese genres including forms of domestic fiction (*katei shōsetsu*). Girls' fiction (*shōjo shōsetsu*), which still affects important cultural products such as Japanese comics and animation, could count Yoshiya among its originators. Yoshiya herself seems to have been well read and affected not only by girls' fiction from abroad such as Frances Hodgson Burnett's *The Little Princess*, but also by a wider reading of European fiction, the decadents in particular.[6] She brought this rich background to girls' culture.

Finally, her relationship with Monma Chiyo, along with the depiction of same-sex relationships in her work, also makes her important to understanding cultures of sexuality in Japan. Any comparative history of lesbian culture should also include her. That this public persona and her fiction that overtly discounted heterosexual norms became a major part of mainstream popular culture is of particular interest in considering the history of gender and sexuality in Japanese culture.

Yoshiya's same-sex relationship is a major reason she might be seen as a "bad girl." But she was almost equally "bad" in the way that she depicted heterosexual eroticism in some of her domestic fiction; often it was explicit enough to require her or her editor to convert long passages to x's or other blank marks to avoid censorship.[7] One wartime work *A New Day* (Atarashiki hi) was taken out of *Osaka Asahi Newspaper* in 1942 for the same reason.[8]

In terms of her public image as well, she was a bad girl. Despite the feminine flourishes of her writing, Yoshiya's Western fashions, unmarried status, and financial independence made her suspect to some. Other writers even found her large salary troubling—especially the fact that she earned more than members of the Japanese Diet. In addition, from the perspective of "serious literature" her very popularity made her a bad girl. Yoshiya sported various Western styles that have been associated with bad girls. She was interested in architecture and generally lived in Western style houses. She also wore dresses rather than kimono and was one of the earliest public figures to wear her hair short.[9] Bobbed hair was often criticized in the media as representing various forms of dangerous modernization, unfeminine appearance, and uncontrolled behavior. Such hair was linked with the "modern girl," a flapper-like figure and focus of journalistic writings in the 1920s. Yoshiya, like many women of the period, would have distanced herself from the modern girl because of the coquettishness associated with the image, and Yoshiya's particular haircut was a bit different. Still, the short haircut marked her as modern, and the outlawing of short hair on women, while rarely enforced, suggests the danger associated with that style choice.[10] Yoshiya herself rarely asserted political meanings for her hairstyle, but defended it against critics. She mocked newspapers' attempts to explain it in a social-scientific way, noting that women probably chose the style because they "happened to have a nice cloche hat" ill suited to Japanese hairstyles.[11]

Shōjo Good and Bad

The most interesting aspect of Yoshiya's work for thinking about "badness" was the *shōjo* (girl) category, which I will look at as found in her 1920 novel *Two Virgins in the Attic*. Yoshiya's first publications were reader submissions to magazines such as *Girls' World* (Shōjo sekai). She recalls that this was the first type of magazine she could have to herself rather than sharing with her brothers.[12] Her reader submissions, followed by her series of *Flower Stories*, make her one of the founders of *shōjo* culture. The *shōjo* past and present has attracted academic and journalistic interest for a long time, and is discussed in other chapters in this volume. The term *shōjo* came into common use in the beginning of the 1900s and coincided with Yoshiya's own girlhood. Like many categories of this sort, the *shōjo* always seems larger than life, representing a conflicting set of ideas and anxieties about gender, sexuality, consumption, education, and Japanese culture. As such, she has always had a complicated connection to the "bad" and the "good." Paradoxically, even though the *shōjo* embodies a hyper-feminine ideal, she also poses an ominous threat to the feminine sphere.

The early image of the *shōjo* was threatening because she was outside the family system; she inhabited a liminal space between the close supervision of her parents and that of a husband after marriage. This status was enabled by modern life, particularly the development of education for girls: it is not surprising that much *shōjo* fiction, including Yoshiya's, takes place in girls' dormitories. As teenagers and young women moved to the cities to pursue new jobs or advanced training for careers such as teaching or nursing, they often had an extended (or permanent) period of life away from family control, an extended time in the realm of the *shōjo*. She thus was seen as vaguely dangerous or corruptible, as indicated by the interest in "degenerate schoolgirls" in the Meiji period, as Melanie Czarnecki's chapter in this volume describes, or in "delinquent girls" (literally, "bad girls" *furyō shōjo*) in the 1920s. The public imagined that the *shōjo*'s unregulated behavior was leaving her open to promiscuity and petty theft, or simply the failure to continue to the next stage of life of respectable marriage.

At the same time, the *shōjo*'s relationship to the school was a source of her potential goodness. Much turn-of-the-century education for women emphasized becoming good wives, wise mothers. The school was seen as the best location to do this well, replacing superstition with scientific and nationally unified standards of behavior for girls and women. Girls having romances and crushes among themselves were seen by some as a positive training ground for future kindnesses to husbands and children. At any rate, educators believed that the *shōjo* was easy to indoctrinate, being safely under the control of modern institutions such as public schools, missionary schools, and dormitories. Christian reformers and other educational experts focused much attention on schoolgirls as a way to improve the nation. To this day, the image of a sailor-suited schoolgirl cheering her team at the national high school baseball tournament remains available to the Ministry of Education as a symbol of a wholesome Japan, just as easily as it might be found in animated shows or movies that are less wholesome.

More abstractly, the word *shōjo*, as Jennifer Robertson provocatively puts it, means "literally, a 'not-quite-female' female"; both her gender and sexuality are ambiguous and flexible.[13] It was in part in her social role that she was not fully female: although identified as a girl, she was neither fully daughter nor wife. In addition, her unformed sexuality meant that she might engage in romances with other girls, in relationships referred to by terms such as "S" (which can refer to "sister," *shōjo*, or the German *schöne*) and *odea* (dear).[14] The history of the reception of such relationships is complex. They were condoned by many as a natural stage in life, innocent and commonplace. Many thought them to be non-physical, platonic relationships and thus unthreatening. However, growing interest in sexology that defined "normal" sexuality and some high profile cases of double suicide increased negative portrayals of such "thickened friendships" and provoked alarm.[15]

The flexibility of gender and sexuality did allow the *shōjo* to be seen as sexually inexperienced and pure, if sexually unregulated. As Robertson notes, the term "implied heterosexual inexperience and homosexual experience."[16] Heterosexual innocence was a common image of the *shōjo* by the 1920s when she was associated with "purity" and warmth. More generally, the *shōjo* often was defined in literature and art by qualities associated with femininity at the time—sentimentality, interest in flowers, clothing, dolls, and dreamy thoughts of the moon and stars. Such *shōjo* are the most sympathetic characters when they are kind and emotionally sensitive or considerate of others.

These last images of the dreamy girl surrounded by flowers and moonlight dominate Yoshiya's girls' fiction. At times her characters seem almost too good, and too feminine, to be interesting. Like much of *shōjo* culture, the fictional settings inside dormitories and girls' homes, and thus away from general society and immune to the outside world, also makes her work seem escapist.[17] On the one hand, this is a rather simplistic view of the political nature of childhood and its relationship to gender. Still, this criticism of escapism reminds us that few adolescents of Yoshiya's day had the leisure to be *shōjo*, except through fiction, as they were much more likely to be working on farms or in textile factories than living in dormitories. Other women writers, Yoshiya's contemporaries, took no such romantic view of girlhood, but preferred to draw attention to the brutal working conditions and gender discrimination faced by working-class girls the same age as Yoshiya's *shōjo* characters. *Shōjo* meanwhile, was in large part a commercially defined image created through products such as *shōjo* magazines, and this is worth keeping in mind in evaluating Yoshiya's girls' fiction.

However, the *shōjo* subjectivity enabled many forms of interpretation and performance that both reaffirmed and subverted mass produced normative ideas of girlhood and womanhood. Yoshiya's fiction permits a particularly wide range of readings of gender roles, as we see in *Two Virgins in the Attic*.

Two Virgins in the Attic

Two Virgins in the Attic begins as the heroine, a schoolteacher-in-training Takimoto Akiko, is announcing to her dorm mistress that she will move out, having been unable to resolve various doubts about her academic motivation and Christian faith. Most of the novel takes place at the YWCA ("YWA" in the Japanese) dormitory to which she relocates. The YWA assigns her the only vacancy, a strange "triangular blue room" in the attic. Akiko's first night nearly becomes her last when she carelessly sets her own quilt on fire with a paper lantern. Luckily, her neighbor Miss Akitsu saves her,[18] and this is the beginning of the friendship between the two women, which develops into a romantic relationship. Akiko is a timid and dreamy young woman, alienated from her surroundings and who, excelling neither in school nor prayer and having lost both of her parents, lacks confidence and is quite miserable; Miss Akitsu, a more spirited and rebellious woman, helps draw Akiko out of her shell. Such a pair—a shy, insecure girl and a spunky friend—remains a popular combination in contemporary *shōjo manga* and *anime*. Other major characters in the novel include Kudō Ryūko, a feminist thought by some to be modeled on Hiratsuka Raichō (1886–1971), and Ban Kinu, a teacher who also has had a relationship with Miss Akitsu and wishes to renew it.[19] After the tension produced by this love triangle and the tragic death of Kudō from illness just before her wedding, Akitsu and Akiko leave the YWA together and find "their own path" together outside the protective attic.

Although this novel is not autobiographical fiction, the situation does parallel Yoshiya's own life in dorms in Tokyo. She also went to live in a YWCA dorm after being expelled from her American Baptist dormitory, apparently for missing both a lecture by Nitobe Inazo (1862–1933) and her curfew while attending movies in the Asakusa theater district. At the YWCA in Kanda, she had a roommate from

Tsuda Women's College named Kikuchi Yukie—thought by many to be the model for Miss Akitsu, with Akiko being based on Yoshiya.

The main characters in *Two Virgins in the Attic* are not what we would usually think of as girls. Akiko is nineteen or twenty years old, and Miss Akitsu slightly older. But there are several ties to *shōjo* fiction. Yoshiya's contemporary readers expected writing in that genre because Yoshiya was already so well known for her writings about girls, *Flower Stories*, and the girls' dormitory setting was the quintessential locale for *shōjo* fiction. More importantly, the writing style, including its florid use of adjectives, romantic imagery, strong emotional expression, and diacritical marks such as exclamation points, all marked it as girls' fiction.

Consider the scene in which an angelic Miss Akitsu saves Akitsu from the fire:

> With each passing moment the flames only spread wider, thick white smoke curled around the room, Akiko lost her breath and was terribly exhausted, and the best she could do was burn her own lips while trying futilely to blow it out! (Ah, what a helpless, pathetic dimwit she was!) And then Akiko's tears began to fall.
>
> Tears, tears—they flowed endlessly. (Oh! If only those tears could have fallen in such as way as to extinguish the fire!) Her voice, choked with smoke, cried out in the darkness.
>
> The door squeaked open.
>
> A human figure stood there quietly—beyond the flames, the gentle sleeves and gathers of the white nightclothes on that person's body were lit up by the scarlet flames, like peaks capped with pure white snow that were tinged by the morning light. That person, like a white crane turning its wings to fly away, swiftly retreated beyond the door.[20]

This passage contains classic images of early girls' fiction: tears, sunlight, birds, and feminine clothing. The narrative expresses strong emotions and self-doubt using rhetorical devices: interjections (*aa*, *ee*), exclamation points, dashes, and repetition. Frequent line breaks provide a sense of importance to individual statements and a sense of drama, while they also inject the feel of poetic discourse into the prose. Yoshiya used this form in her nostalgic preface to *Flower Stories*, to address her *shōjo* audience as a former *shōjo*:

> My girlhood days that will never return
> Those flowers that blossomed in my dreams
> I send every lovely one
> To all of you[21]

Many critics have viewed the characters (or in the case of the preface, her own authorial persona) who are depicted using this style as narcissistic and self-absorbed. Even Yoshiya herself was mildly critical of the lack of character development and cliché imagery that such writing exhibited. She reflects back on her early days as a writer when she would submit essays and stories to girls' magazines as "reader submissions":

> The style of those readers' submissions was *bibun* (ornate) (e.g. "When I happened to look up at the sky, I saw the faintly moonlit sky and was overcome by tears"). This

sort of writing had taken the world by storm and no matter who wrote a given piece, there would be darkness, looking up at the sky to see moonlit skies, the evening star, or evening clouds, and then, for no reason, a bursting into tears. Many girls became well known for this sort of submission at that time.[22]

These reflections might suggest that all little girls represented would react identically to the same trite images, as Yoshiya seems to envelop all female characters in the same pretty imagery and an unchanging emotional state. I would like to argue, however, that her writing allowed girls and women multiple forms of identification.

To this end, first we can look to the ways that this excess in writing and emotions plays out in both style and plot. Just as Yoshiya's reflections on *bibun* above suggest, the artificiality and predictability in this style is obvious even to its authors. In *Two Virgins in the Attic*, this mannered way of writing calls attention to language itself; in a famous passage Akiko considers the marvelous "blue triangular room" and wonders how to capture it in language:

What name could she use that would encompass all of that? A word to articulate that strange world! She grasped for a word. Akiko again looked up to the rafters, higher and higher. And the next instant—a sign crossed her mind, from out of the many words in her vocabulary, as if it had been safely hidden away just for today, just until now—like a golden key that suddenly turned up in a useless old box of toys, that one word—letters appeared to form it, as if written across the sky. Clearly and distinctly—. . . . . ATTIC . . . . . . ATTIC . . . . . . ATTIC![23]

Through this dramatic discovery, "attic" expresses more than the architectural space; it has strong visual and emotional impact, first because of the roman capital letters (ATTIC) and exclamation points, and second, because it is revealed through the same trope seen throughout girls' fiction of a young woman looking up to the sky. Here, instead of seeing a starlit night or moon (although she sees plenty of those elsewhere in the novel) she catches sight of an English word "written across the sky." This passage presents the word "attic" as significant and says that the space and the word must incorporate a broad range of meanings, which will in turn allow its inhabitant to find a different way to live. The attic lies outside of the Japanese family structure or other institutions (school, dormitory administration, church), and is even not quite Japanese; this is what will permit Akiko to explore different options for living as a young woman in the world. The association of the attic with foreign lands, made richer and visually striking through use of English orthography, produces a number of associations that stretch the possibilities for this young woman from the Japanese countryside.

She had heard that these were the preferred abodes of people far across the sea who lived a life of poverty eating salt water and black bread—and yet, and yet, here in the Oriental island country of cherry blossoms—in the land of a gentle people who still used stone lanterns and stepping stones—did they not in fact need that seemingly unnecessary noun born from the English dictionary: "*yaneura*." And so, one young girl of this country, came to live in that little safe domain![24]

Here Yoshiya contrasts an exaggerated Japanese cultural identity to the attic, which she links linguistically and thematically to Western culture and a bohemian artistic existence (the term atelier is used elsewhere in the novel as well). This sort of exoticism was common in *shōjo* fiction and also contributed again to the sense that girls in the stories were living in a larger world, one that reached beyond the confines of family, school, or even nation. The humble Akiko reads as both adorably small and needy in this "little safe domain" and part of something larger and more important. Throughout *Two Virgins in the Attic*, Yoshiya uses Western images of architecture (attics, balconies, music halls), furniture (pianos, tables, chairs), and people (Romain Rolland, Oscar Wilde, Annette Kellerman, St. Francis).[25] Moreover, Christianity and the non-Japanese missionaries who surround her give a sense of foreignness within Japan, exoticizing both the West and East while providing room for imagination and identification beyond a narrowly Japanese girlhood and womanhood, opening up a cultural world beyond boundaries of gender, sexuality, and nationality.

Interestingly, Yoshiya herself lived out this interest in the foreign and its possibilities. Upon receiving her first large royalty payment in the mid-1920s, Yoshiya traveled to Paris and later turned her experiences into the serialized work *The Stormy Rose* (Arashi no bara, 1930–1931). She also traveled around Asia during the early years of World War II; published books and newspaper essays tell of her visits to China, Manchuria, Thailand, Indonesia, and Vietnam.[26] Again, this type of experience was not available to most readers, but the image of the young woman able to fund her own trips to Europe without the aid of family money or a husband was inspiring to others. Within her novel as well, different spaces in the dorm allow for experiences of her good girl and bad girl self.

The attic in many works of girls' fiction has been a place of loneliness and deprivation, as in *The Little Princess*. For *Two Virgins in the Attic*, however, the attic provides privacy, space for reflection, and tender friendships not permitted in other spaces that required adherence to "good girl" expectations. Rather than seeking to move down to the regular rooms below, Akiko and Miss Akitsu look upwards (in the romantic *shōjo* mode to the sky, stars, and birds) or, finally, outside the dormitory:

(Our attic)
Let's say it clearly—ah, it really was (Our attic)!
(Our attic)
It's now time to say farewell—
Those two young women who could live together in that little beloved triangular blue room leave it behind—
The two of them take their respective fates as handed to them—
And so, now the two of them seek out the paths that they will follow and leave the attic's blue room.
(Our attic)
Farewell!
The attic, the beautiful wine bottle in which their (fates) have aged.
The blue cradle of an attic that nurtured the two maidens' (fates).

Farewell—
So, the two virgins Akitsu Tamaki and Takimoto Akiko placed on the attic's blue walls a
parting kiss, and left together.
Seeking a new fate!
Searching for the paths they should follow![27]

Having left their "Starting Point" (written in English) of the attic, they build a new
life together rather than returning to their original "fates."

As Hiromi Tsuchiya Dollase has emphasized, Yoshiya's girl fiction is very much
concerned with fantasy and creativity, and the attic becomes a space for that as
well.[28] Akiko often indulges in a reverie where she sees herself, for example, performing Beethoven on the piano for a large audience of foreign ladies and gentlemen; or
as a dragon-slayer going up into the tower of a castle. These fantasies operate not
unlike the images of foreignness to broaden the world of the attic and girlhood
itself. The attic becomes a space for creative play and art that allows Akiko and her
friends to think beyond themselves. The spunky feminist friend Miss Kudō forms
"The Apple Association" (also known as "The Black Hand" society) which meets in
the attic. A vaguely political group who eat apples together, its meetings provide an
alternative to the church services and school lessons, and a connection to the world
outside. In a simpler sense, Miss Akitsu produces various works of art in the attic,
most playfully a horse sculpture that contains an apple (until someone eats it) and
a ruby ring. One of the Apple Association members worries that Miss Akitsu's ruby
ring will rust while in this sculpture, but for most the attic is a space beyond such
practical concerns of family prestige and wealth; it is a refuge from concerns over
whether Miss Akitsu or other girls will be expected to marry.

Finally, the attic and other Western spaces of the novel provide a space for sexual encounters and unions different from those readers might see in the moving
pictures or read of in other novels, even those that use the same set of girls' fiction
imagery. The balcony provides the first such space, where the two girls meet after
taking their baths in the dorm:

"Balcony"—She was so happy that there was such a wonderful thing. After going
down that wide corridor on the third floor and off to the right there was a big glass
door. Miss Akitsu opened it and the two of them went outside—balcony—. . . .
Turning, she saw Miss Akitsu standing there silently in her beauty, rapturously grasping the railings, playfully fingering the black hair at their shoulders, her eyes clear
and bright, catching the fluid moonlight; what were her thoughts? . . . If a mermaid
were deep in lamentation as she sat upon a rock on the shore, gazing longingly at the
moonlight she would look like this . . . . . . Miss Akitsu placed on her palm a small
glass bottle filled with amber liquid, and squeezed the rubber globe at the top to
release a mist in the moonlight . . . . . . the mist of perfume took on the faint colors
of the rainbow and disappeared behind the two maidens' hair . . . . . .
    Forever, forever, forever, if they could stay like this—even if the earth was going be
destroyed someday, at least until then—this is what Akiko wished for in her heart.[29]

The dreamy language, foreign architecture, and art nouveau-like scene—all
features familiar to girls' fiction readers—are also what allow for the passionate
expression of Akiko's feelings for her dorm-mate. Here the space seems a fantasy

space, with mermaids and moonlight, but it also reflects on reality: the girls will be expected to part when they grow up rather than staying together "forever." Thus when they finally do leave the dorm together, readers are reminded of the social obstacles they are overcoming to be together. Although there are several scenes along the same lines, the most vivid description of the two girls' physical relationship is of the way they share a bed, having made their two attic rooms into one:

> In this cramped, blue, triangular room that one bed of such rare beauty was enough. And so that one beautiful place to sleep was all the two used night after night.
> . . . Miss Akitsu's linen pajamas have the soft scent of magnolias . . . . . . and unnoticed that scent of the magnolia flower was transferred to the flannel sleeves of Akiko's own sleepwear . . . . . . so like a magnolia to slip its fragrance into the bedroom during the night . . . . . . their arms were resting as if entwined . . . . . . the hearts in each of their breasts softly ticking . . . . . . as if their two souls had disappeared into a tender dream without beginning, and without end . . . . . . a soft, pliant kiss . . . . . . a kiss like trembling and melting into a damp, red petal . . . . . . softly, tenderly flowing, sinking and surfacing, disappearing and melting into the replete then ebbing undulation . . . .[30]

Again, the young women's sexuality is expressed in the girls' fiction language of flowers, making it familiar but open to new interpretations. Although many journalists and thinkers condoned same-sex romances among girls, *Two Virgins in the Attic* here pushes the limits of its contemporary *shōjo* fiction. Girls' romances were assumed to be physically distant if emotionally intense; Akiko and Miss Akitsu's relationship has also been described that way by readers even though the text suggests otherwise. The confusion comes in part because the love is described as "pure," but this should not be taken to mean it is not erotic. In fact, their physical connection is initiated through Akiko's own purity. Akiko has honestly expressed her doubt about the existence of God at a church meeting. When Akiko confides that she wishes God would appear before her, the other girls laugh at her naiveté. Miss Akitsu comes to her after the meeting:

> . . . . . . behind her Akitsu felt the warmth of another's flesh closing in on her . . . . . . a gentle arm softly and warmly embraced Akiko's trembling shoulders . . . . . . hurried, excited breaths rushed across Akiko's cheeks . . . "You . . . . . . You . . . . . . are, what can I say . . . . . . a pure, honest . . . . . . person. . . ". . . As rough and wild storm rains buffeted the dark building, the spattering rain streamed down the peaked roof like a waterfall, reverberating as its shook the attic bed.[31]

Perhaps this focus on "purity" is what has allowed even this relationship, which is clearly a quite sexy one, to be read as innocent. Alternatively, for those who seek a strong lesbian novel, this might be seen as insufficiently explicit, a sort of immature love between girls rather than between adult women.

Instead, I would argue that Yoshiya's focus on what is generally viewed as a transitional moment between girlhood and womanhood, makes us rethink those ideas about the nature of that move and its relationship to notions of femininity, as will be explored in the next section. A look at the theme of growing up in

Two Virgins in the Attic and Yoshiya's thinking will help clarify the cultural significance of good and bad *shōjo*.

Growing Up Good and Bad

Early in the novel, Akiko resists the social pressure to move into adulthood, escaping into readings about saints:

> As her girlhood days went by, Akiko held her breath upon reading St. Clara's autobi-ography and stories of St. Francis, and constructed out of them a deep, deep world of adoration. When she tried to pass from those girlhood days into the realm of adult womanhood, that beautiful illusion, began, if not to break, at least to dissolve. Not only was that illusion fading away, but also there was no sign of something being built to replace it. Like a demolition without any construction to follow, her spirit was a cavity holding nothing but tears of anxious gloom.[32]

This novel has rightfully been read by Michiko Suzuki as a *bildungsroman* or grow-ing up novel that strives to depict "same-sex love as neither deviance nor sexual innocence."[33] Akiko has an identity crisis: she sees her *shōjo* life crumbling around her. Miss Akitsu's love helps her find something to replace it, and together they move beyond the attic into adulthood. Equally importantly, Akiko's growing up does *not* entail heterosexual marriage or loss of purity. In fact, it is quite important that the concept of "purity" here is separate from the issue of sexual experience and does not imply that she is innocent of erotic sensations. Akiko has already moved beyond sexual innocence through her relationship with Miss Akitsu, so the drama of the novel's conclusion has nothing to do with its loss. She also does not leave behind the "purity" associated with girlhood when she leaves the attic with Miss Akitsu. Yoshiya does not represent this as a failure to mature fully, some sort of infantilized adulthood, but instead a self-affirming choice to remain "pure."

Close to the time that Yoshiya wrote *Two Virgins in the Attic*, she was beginning to have doubts about the world of *shōjo* publishing, which increasingly narrowed the possibilities for the meaning of girlhood. In a private letter to Monma Chiyo a few years later:

> The girls' magazines are so terrible lately that I just sigh every time I read them . . . Almost to the point of endorsing obscenity, they push on girls the idea that they should be flirting with men. Are the artists and writers who publish there really human? They do this even though it is true that anyone having a pure spirit, anyone who is simply given the opportunity to polish it can enjoy her solitude; still these people take their filthy hands and cover girls' eyes, propping them up as so many clay dolls that can only think of marriage—they are truly a dirty bunch. With *Black Rose* [her self-published magazine] I raise a protest banner against that trend; Nobuko will write, throwing flower petal after petal against it. To guide the *shōjo* out of that darkness, I will lead them with my own hands out onto a lighted path. Like Joan of Arc, I will wield *Black Rose* like a sword against the male writers who lead them astray. I will do battle with them face-to-face shouting "begone you demons" and exorcise them from our midst.[34]

Yoshiya criticizes mainstream *shōjo* culture for its emphasis on marriage and coquetry, and uses *shōjo* flower petals to do so. Yet her reference to Joan of Arc, sword fighting, and a battle against evil, transforms those petals into swords. This is typical of Yoshiya's use of imagery commonly associated with girls in combination with other less accepted images of women such as Joan of Arc to build a wider range of options for girls and women that transform them from "clay dolls" to something more interesting. Similarly, the ending of *Two Virgins in the Attic* maintains the language and sincerity that she has associated with girlhood, but transmutes the girls into willful selves rather than good doll-like figures.

In a forcefully argued attack in *Black Rose* on writer Morita Sōhei (1881–1949), who represented chastity negatively in his novel *Transmigration* (Rin'ne, serialized in 1923–1924 in the magazine *Josei*), Yoshiya explicates her own theory of gender in terms of chastity and purity:

> Therefore, just as there are men who refuse the fate (?) of being husbands or fathers, so too must women refuse to let people take away all options besides being wives. If that is considered rebellious, I say what is wrong with that—is not the basic daily life of humans an ongoing rebellion against nature? The building of civilization is a record of that rebellion. There is truth to the argument that males have gained what is good and whatever things make them great through struggle . . . Nature (or the instincts of one's species) manipulates like a puppeteer, whether one is male or female; nature does not recognize personality and does not distinguish one individual from another. However, you might say that in some sense nature has been especially harsh on women and has exploited her efforts. With the heavy burden that nature has placed on women, creating and nurturing some sort of individuality requires an even greater degree of rebellion against nature, a greater struggle. And no one should denounce that rebellion, it is progressive and constructive. Her purity does not come from ignorance but from wisdom. Her chastity is not forced on her by a proprietor but is something that she desires from within her self. It is something chosen not out as the naïve happiness of a pig, but out of the great struggles and tragedies that characterize human life alone.[35]

"Purity" and "chastity" become not some ideal natural aspects of the female sex, but rather something women, both individually and as a cultural unit, construct as a positive quality, one that is not conservative but in fact rebellious. By rebelling against natural imperatives to marry men and reproduce, she becomes a more civilized human (as opposed to the more natural pig) capable of developing an individual personality and a constructive place in society. In this way, "purity" is reworked into a form of civilization and maturity rather than the incompleteness of a girl. Akiko's purity and honesty need not be discarded in adulthood, but instead become the tools that enable her to act in a way that is true to her feelings, innocent or not. More broadly, Akiko and Miss Akitsu's relationship becomes a normal rebellion against nature rather than a form of deviance. Their rebellion against expectations for their biological sex opens up other possibilities for defiance of nature, even paths other than those which these "two virgins" follow. The bad girl is the good girl.

Two Virgins in the Attic was Yoshiya Nobuko's last major novel to focus explicitly on a lesbian relationship, and she did not write one about the type of life she lived with her partner Monma. Such a novel might have been moving and interesting, but perhaps it was because Yoshiya went on to write domestic romances that included wives and mothers as major characters who drew a large audience that she was able to live a peaceful life with Monma in financial security and independence. Her later novels, which are certainly worthy of study, also recognize that the real worlds of most of her readers would more often parallel a domestic romance novel than a lesbian story and that few readers would have had the option to live as she did if they wanted to. Instead, Yoshiya brought the same celebration of feminine virtues seen in girls' fiction: kindness, "chastity," "purity," and depth of emotions, to characters such as mothers, widows, and sisters. Rather than allowing this choice to narrow her depiction of women, she continued to write such stories in a way that lent flexibility and interiority to female characters.

Conclusion

The title of this article contains a tongue-in-cheek reference to Britney Spears' song lyrics ("Oops, I did it again," 2000) that is worth turning to briefly. Spears' growing up story has been a matter of public interest derived largely from voyeuristic pleasure taken in a crumbling girlish Mouseketeer image, chipped away at by such events as kissing Madonna and a twenty-four-hour long marriage, as well as the fashion choices made for her by her stylists—all of which suggested that she was "not that innocent." These events attract attention largely because Spears is a *shōjo*-like figure, a not-quite-woman who was both girlishly innocent and sexy; her ambiguous status as a publicly declared virgin also came into play as well. She has the commercial appeal of being a good girl gone bad.

The *Two Virgins in the Attic* gives us a very different view of the girl in transition. By having a close examination of the interior world of Akiko as a thoughtful and sensitive girl, Yoshiya never lets the reader think that she is "innocent" or even wants to be, since we see right inside her love affair, leaving the reader utterly uninterested in watching her fall from grace. This succeeds in making her virginity a matter divorced from her erotic life, because for her sex with a man would in fact be destructive to her erotic desires. "Virginity" signifies the two girls' escape from the social position of wife, not denial of their sexual desire. Instead, the drama in Akiko's transition from girlhood to womanhood is found in her search for self and in reader anticipation about Miss Akitsu's decision to stay with her. These events are expressed not through a transformation to adulthood, but through the language of girl culture that remains usefully flexible to them well beyond adolescence. While some readers would not, of course, have been interested in the lifestyle that Akiko (or Yoshiya) pursued, and others would not have access to it, the detail and creativity with which Yoshiya depicts girls growing up provided a wide range of ways that readers could understand their own relationships to notions of femininity and sexuality that surrounded them. Perhaps it is not surprising that a sizable group of Britney's North American contemporaries have turned their interest to the Yoshiya-influenced world of *shōjo anime* for a different depiction of gender and girlhood.[36]

Notes

1. Yoshiya Nobuko, "Foxfire" (Onibi), trans. Lawrence Rogers, *The East* 36, no. 1 (May–June 2000): 41–43.
2. For example Yoshiya Nobuko, *Zuihitsu: Watashi no mita bijintachi* (Tokyo: Yomiuri Shimbunsha, 1969) and *Jidenteki joryū bundanshi* (Tokyo: Chūō Kōronsha, 1962).
3. Tanabe Seiko, *Yume haruka Yoshiya Nobuko: Akitomoshi tsukue no ue no ikusanka.* 2 vols. 1999. Reprinted (Tokyo: Asahi Bunko, 2002); Yoshiya Nobuko, *Otome shōsetsu korekushon* (Tokyo: Kokushokan, 2000–2003). *Wasurenagusa* is daylily, but its literal meaning is closer to "forget-me-not."
4. Kawamoto Saburō, "Jiritsu suru 'onna Yoshikawa Eiji,' " *Shokun!* (January 2003): 276.
5. Tanabe Seiko, *Yume haruka* 2: 682–683.
6. Decadence was not a formal literary movement but is a label placed on a number of writers of late nineteenth century France who often focused on degeneration and deviance.
7. Sarah Frederick, "Sisters and Lovers: Women Magazine Readers and Sexuality in Yoshiya Nobuko's Romance Fiction," *AJLS Proceedings* 5 (Summer 1999): 311–320.
8. See Jennifer Robertson, "Yoshiya Nobuko: Out and Outspoken in Practice and Prose," in *The Human Tradition in Modern Japan*, ed. Anne Walthall (Wilmington, DE: Scholarly Resources, 2001), 169.
9. Barbara Sato, *The New Japanese Woman: Modernity, Media, and Women in Interwar Japan* (Durham, NC: Duke University Press, 2003), 51–56.
10. Robertson, "Out and Outspoken," 162.
11. Yoshiya Nobuko, "Danpatsu oshikari no koto," *Kuroshōbi* 3 (March 1925): 52–56.
12. Yoshiya Nobuko, "Tōsho jidai," in *Yoshiya Nobuko: Sakka no jiden*, vol. 66 (Tokyo: Nihon Tosho Sentā, 1998), 33–34.
13. Jennifer Robertson, *Takarazuka: Sexual Politics and Popular Culture in Modern Japan* (Berkeley: University of California Press, 1998), 65.
14. For detail in English see Robertson, *Takarazuka*, 68; Michiko Suzuki, "Developing the Female Self: Same-Sex Love, Love Marriage and Maternal Love in Modern Japanese Literature, 1910–39" (Ph.D. diss., Stanford University, 2002), 28–39.
15. See Sabine Frühstück, *Colonizing Sex: Sexology and Social Control in Modern Japan* (Berkeley: University of California Press, 2003).
16. Robertson, "Out and Outspoken," 159.
17. For example see Satō Tsūga in Hayashi Sawako, "Toshokan shiryō to shite no taishū jidōbungaku o kangaeru: Yoshiya Nobuko no shōjo shōsetsu o rei ni," *Ōtani Joshidaigaku kiyō* 35, no. 1.2 (March 2001): 189.
18. Miss Akitsu is called by her family name throughout (*Akitsu-san*). Akiko is a given name, and this is an unusual way to call a female character in modern Japanese novels. Unfortunately, English makes it difficult to convey the more gender neutral feel of Akitsu-san.
19. Yoshikawa Toyoko, "Seitō kara 'taishū shōsetsu' e no michi: Yoshiya Nobuko Yaneura no nishojo," in *Feminizumu hihyō e no shōtai*, ed. Iwabuchi Hiroko et al. (Tokyo: Gakugei Shorin, 1995), 132.

20. Yoshiya Nobuko, *Yaneura no nishojo* (Tokyo: Kokushokan, 2003), 79–80. (Original publication. Tokyo: Rakuyōdō, 1920.)
21. Yoshiya Nobuko, *Hana monogatari*, 2nd edition (Tokyo: Kokushokan, 1995), unpaginated preface. Original publication, 1920.
22. Yoshiya, "Tōsho jidai," 35.
23. Yoshiya, *Yaneura*, 50–51. Ellipses and dashes in original. Yoshiya uses ellipses in six dot increments, which I will preserve. For my own ellipses, I use the standard formatting, (. . .). "ATTIC" is written in roman letters in the original.
24. Ibid., 51. She uses Chinese characters with the *kana* gloss "*yaneura.*"
25. Romain Rolland (1866–1944), French author and activist, well-known in Japan; Annette Kellerman (1887–1975) was an Australian movie actress.
26. See my "Bringing the Colonies 'Home': Yoshiya Nobuko's Popular Fiction and Imperial Japan." In *Across Time & Genre*, ed. Janice Brown and Sonja Arntzen (Department of East Asian Studies, Edmonton: University of Alberta, 2002), 61–64.
27. Yoshiya, *Yaneura*, 317–318. Parentheses in the original and used for emphasis of particular words.
28. Hiromi Tsuchiya Dollase, "Early Twentieth Century Japanese Girls' Magazine Stories: Examining *Shōjo* Voice in *Hanamonogatari* (Flower Tales)," *Journal of Popular Culture* 36, no. 4 (Spring 2003): 724–755.
29. Yoshiya, *Yaneura*, 153–155.
30. Ibid., 195. Final ellipsis is doubled in length for emphasis.
31. Ibid., 188. Six dot ellipses are in the original. Regular ellipses are mine.
32. Ibid., 170.
33. Suzuki, "Developing the Female Self," 27.
34. Yoshitake Teruko, *Nyonin Yoshiya Nobuko* (Tokyo: Bungei Shunjū, 1986), 32–33.
35. Yoshiya Nobuko, "Haru o mukae no bundan susuharai," *Kuroshōbi* no. 1 (January 1925), 64–65. Question mark in original.
36. *Revolutionary Girl UTENA* (*Shōjo kakumei utena*, TV series April 2–December 24, 1997; movie 1999) makes direct reference to *Two Virgins in the Attic*.

The file photo of Abe Sada taken when she was 31 years old. The photo was taken May 19, 1936 (Kyoda). Tokyo, Japan.

5

So Bad She's Good: The Masochist's Heroine in Postwar Japan, Abe Sada

Christine Marran

The Abe Sada Incident

The week of the murder of Yoshida Kichizō by his lover Abe Sada was one filled with passion. The two had been meeting secretly for months behind the back of Kichizō's wife who was also Sada's boss. This time the lovers snuck off to the Masaki Inn on May 11, 1936. Sada remembered in testimony the passion of their lovemaking in reflecting on one tryst at the Tagawa inn:

> We kept the bed out from the evening of the 27th to the morning of the 29th, and hardly slept at night doing every nasty deed possible. When I said I was tired Ishida would make love to me and even while sleeping he would massage my body very sweetly. It was the first time in my life that I had met a man who treated a woman so well and who made me so happy. I fell in love. I could never be separated from him. . . .[1]

The couple experimented sexually. Sada spoke in testimony about the sex play they engaged in prior to the murder:

> The evening of May 16th I got on top of Ishida and at first we had sex while I pressed his throat with my hands but that didn't do anything for me so I wrapped my kimono sash around Ishida's neck and I pulled it tight and then loosened it and so on while we were having sex, and while I was doing it I kept looking down there so I didn't realize that I'd squeezed too hard; Ishida let out a moan and suddenly his thing got small. I was shocked and released the sash but Ishida's face [and neck] had turned red and didn't return to normal so I tried cooling his face by bathing it with water.[2]

pair's relationship lasted more than three months but the nearly fatal choking incident perhaps encouraged Kichizō to rethink his situation because soon after his near asphyxiation, he suggested that he return home to his wife for the time being. This did not mean that Kichizō would break off with Sada, just that he would be more discreet. He suggested that he set Sada up as his mistress. However, when Sada heard that she was to share Kichizō's affections with his wife, she erupted in anger. During the couple's sexual play later that evening, Sada again wrapped Kichizō's neck with the pink sash of her kimono, this time, fatally asphyxiating him. She then used a kitchen knife to sever the penis from his body and carved in his left thigh *Sada Kichi futari*—"Sada and Kichi, the two of us." On the bedding she wrote a similar sentiment in his blood. The choice of phrasing was gruesomely appropriate, for its literal meaning was "Sada Kichi the two who are cut [off from the rest of the world]" (*Sada Kichi futari-kiri*). This concept had a history in the Tokugawa period when used by courtesans who promised their loyalty to patrons by offering a token of their affection which in extreme cases was purported to be their own finger. Courtesans would even tattoo a sign of devotion on their upper arms. Sada reversed this practice of self-mutilation and inscribed Kichizō's body with her love letter instead.

Sada eluded the police by wandering the streets of Tokyo and was finally arrested three days after the incident on May 21, 1936. She was convicted and sentenced to six years in prison. The sentence was commuted, however, in 1940. Sada was given an alias and remained hidden from the public eye until she became incensed by a novel written about her by Kimura Ichirō and emerged from hiding to take him to court for libel. She then appeared in several postwar plays portraying herself. Reflecting on the case in 1947, eleven years after her arrest in 1936, renowned author Sakaguchi Ango declared that he had yet to see a crime gain as much attention in the newspapers. He criticized the melodramatic approach taken by the press, which he felt not only betrayed the feelings of the readers but the reporters as well: "The journalists writing in such a sensationalist tone were actually Sada's greatest sympathizers and empathizers, yet betraying their own hearts, they ardently sensationalized the crime as journalists are sadly wont to do. Indeed, even if there hadn't been the Sada incident, the country was at the gate of an intolerable crushing fascism. Sada too, thanks to the fascist era, was likely made scapegoat in an overly reactionary, sensational ruckus."[3]

The enigmatic Sada had become the new and favored subject of the press soon after her arrest. The newspapers called her "Shōwa's poison-woman," adding her to the ranks of the imaginative tradition of the disorderly woman in Japan. To be called a "poison woman" was to be included among female criminals who were the object of sensationalist attention of journalists and writers and who became the subjects of hundreds of newspaper articles, novels, poems and kabuki plays.[4] Journalistic writings linked Sada to a genealogy in which the woman who committed crimes was lascivious, sexual, and only occasionally an object of sympathy.

Interest in deviant women, especially Sada, continued throughout the twentieth century. Sada's notoriety gave rise to innumerable interpretations of her motive to commit murder. Such interpretations appeared in the form of plays, novels, films, books, articles, and images, making her the most well known bad girl in modern Japanese history. This chapter is dedicated to only a handful of Sada stories and is by no means exhaustive. These particular narratives of Sada have been chosen with the intent of comparing two types of Sada representations: medical diagnoses in which she is depicted as dangerously abnormal sexually and postwar representations of Sada in stories and films that depict her as a sympathetic or even romantic figure and which deviate substantially from the condemning scientific reports made after her arrest. This curious shift of interest whereby Sada is first treated as a medical specimen and then as a heroine in postwar film and literature illustrates the way in which the bad girl can speak to competing interests. The earliest discussions of Sada treat her as a medical curiosity at a time when sexual science is popularized. Given the prevailing interest in sexology at the time of her trial, it is not unusual that one predominate tendency was to treat Sada as a medical specimen and exemplar of the uses to which sexological sciences could be put.[5] Specifically, Sada was accused of exhibiting perversions discussed in contemporary sexual science—fetishism, sadism, masochism, and nymphomania.

In postwar culture, however, Sada emerges as a heroine. In manifold texts ranging from pulp magazine serials of the 1950s to film and theatre of the 1970s, Sada was made to embody a socially subversive position. In other words, Sada was transformed from bad girl to good girl, useful to the attack of the patriarchal status quo, which these writers and filmmakers saw as the real evil. More specifically, this bad girl and her lover were made a foil for critiquing the cultural and political demands of imperialism and in romanticizing the refusal to follow patriarchal imperatives.

The Medical Reports: Crooked Teeth + Flaring Hormones = A Killer?

Controversy over Sada's motives erupted with the first reports of the murder and continued through the trial in 1936. When confronted with the prospect of sharing her married lover with his wife, Sada had refused. Desperate, she sought to possess Kichizō through radical means—an eternally "monogamous" marriage in death. While an extreme act, the "why" of Sada's murder was relatively straightforward. Nevertheless, a variety of hypotheses emerged to explain how she could have done such a thing. Despite Sada's complaints, that the social system in which a man keeping a mistress was acceptable might be such a source of frustration and pain for a woman, was all but ignored in discussions of the case. Instead, Sada's sexual proclivities and openly expressed sexual desire were put on trial.

The police report documenting her case reveals how quickly the police sought to find fault with Sada and how little they explored her claims to rape and enforced prostitution as an adolescent. The police looked to Sada's youth and sexual experience for clues as to why she had murdered her lover:

> She listened to dirty stories of the workers [at her home] and she was influenced by this and matured quickly. In the summer of 1919 she lost her virginity at the

age of fifteen, grew despondent, and looked for friendships amongst the town's delinquents. . . . she prowled around looking for the opposite sex, her licentious lifestyle became so excessive that her father Yasuyoshi thought it in her best interests to make her a geisha [geigi].[6]

The reference to "losing her virginity" had actually been described in testimony by Sada as rape by a Keiō University student. In the police description, however, the rape of Sada at the age of fifteen, the agony of losing her virginity outside of marriage, Sada's forced entrance into prostitution by her father (Sada said in testimony that she cried for three days and begged her father not to send her out to be a sex worker but he was insistent), the various incidents in which she is tricked into greater debt by acquaintance-cum-pimp Inaba Masatake, the catching of syphilis, her attempts to escape locked geisha houses in which she was imprisoned, and her vain attempts to have a sustained relationship are treated as her own failures of character developed in response to early sexual maturation. Nowhere is the father or the student reprimanded. It is Sada who is the sole source of her troubles.

The various medical studies about Sada repeatedly claim a connection between the murder of Kichizō and Sada's reproductive functions and sexual desire. In Japan at this time, physiognomy was considered by some experts as a relevant semiotic index for explaining how Sada could have been driven to commit the crime. The police interrogator addressed precisely this point by asking her whether she had regular periods to which she replied: "I started my period toward the end of my sixteenth year. It's always regular and it's over in about four days. Sometimes I get a headache during my period which makes me crabby but it's not bad enough that I have to lie down."[7] This line of questioning illustrates a continued reliance on biology and female sexuality to understand deviancy among women in the early-twentieth-century Japan.

Sada is matter-of-fact about her period. She maintains that it has little effect on her, but the police report reflects a belief that menstruation might lead to bad behavior, as evident in the following comment: "She began to menstruate at about age fifteen and from that time she continuously quarreled with members of her household." The police were also apparently interested to learn that around the age of seventeen Sada experienced an inflammation of the ovaries.

The connection between Sada's murderous act and her physical health and mental health was made even stronger by a battery of tests she had to undergo. Extended physical and psychological analyses were first performed on Sada for the court. The exams took seventeen days. According to the Japan Psychological Examinations these medical exams were lead by Muramatsu Tsuneo, a professor at the Tokyo Imperial University Medical School.[8] For his report, Muramatsu examined the records of testimony by Sada and her clients, her genetic history, her state of physical and mental health. An almost voyeuristic attention centered on how Sada regarded the male body and how she touched it. Witnesses, presumably former clients of hers, were even brought in to testify. Among the topics discussed in the findings were why Abe Sada considered the penis and the testicles to be "one set," the many men she had been with and what they had to say about her, such as how she was usually more interested in satisfying her own pleasure, how she would lick or suck the penis first sticking it in and then pulling it out, how she would lose her breath after orgasm, or how she was "ten times" more sensitive than usual women to the degree that her eyes would even change color when she was sexually excited. Another witness had said that he feared she was ill because she was wet before sex but she had replied that she had seen a doctor about it and was diagnosed as merely easily excitable.

The state of Sada's current health was deemed part of the evidence as well. Muramatsu's physical exam found Sada to be physically and nutritionally below the norm—her teeth were crooked, and her pupil reaction slow; he wasn't sure whether she had syphilis at the time, but her cerebrospinal fluid appeared "normal." His conclusions regarding her mental state were that while she had tendencies toward sadism and fetishism that she was not necrophilic. She did not gain a sense of pleasure in the act of murder. Surprisingly, she was not found to be pathologically sexually perverse. Her sexual life was unusually active and her orgasms "were unusually long and deep but otherwise qualitatively she does not exhibit perversion that would become pathological."[9] He found that though she was born not breathing, and did not speak until the age of four, this had not made her mentally weak. She grew at a normal rate, her teeth came in at the normal time, and her reflexes were good. Her problem was that she had not been in an environment in which she could correct or reflect upon her immorality and low ethics as a juvenile; rather she had continued to be in places that encouraged them. Syphilis, drug addiction, and alcohol addiction (the amount she consumed before the murder is measured in bottles consumed per day) were found to have not contributed to the incident. However, Muramatsu revised his opinion in view of the court judgment to assert that Abe exhibited mental and physical hysteria which he called nymphomania.[10]

More medical experts joined in the discussions of Sada's mental health and produced many other interpretations. Medical analysis by various members of the medical establishment, however, overlaps little with Muramatsu's analysis. Prominent sexologists and psychologists in Japan including Takahashi Tetsu and Ōtsuki Kenji joined specialists from the Tokyo Psychoanalysis Research Institute and a forensics psychologist from the Tokyo Municipal Police Department to diagnose Sada's behavior. Their results were published in 1937 in *The Psychoanalytic Diagnosis of Abe Sada*.[11] The contributors gleaned much of their information from discussions about Abe run in the widely read *Ladies' Journal (Fujin kōron)*. In this article, former neighbors of Sada discussed her behavior as a little girl (brave and strong-willed).[12] The Tokyo Psychoanalysis Research Institute's purported motivation in publishing the book was to illustrate what normal female sexual development looks like and to promote psychoanalysis as the best model for understanding the impetus to social and sexual deviancy, as exhibited by a woman like Sada. Thus, in their prefatory remarks, the authors state their primary goal to be a will to a truth that can be revealed only through the science of psychoanalysis.[13] They are disdainful of various hypotheses published in newspapers regarding Sada, and are determined to correct the misinformed public. Evidence from the original police report and the state of Sada's health are given less importance here. As the titles of these essays show, Sada's bad girl behavior is thought to have originated deep in her psyche.

The *Psychoanalytic Diagnosis of Abe Sada*, though focused on deviancy, stretched contemporary ideas of what constituted normal female desire. The fact that it did not repeat judgmental platitudes and offered competing diagnoses at least encouraged the questioning of the categories of deviancy. The primary focus of the study is to uncover Sada's impetus to criminal intent through an analysis of her sexual instinct. Within the larger discourse of criminology at this time, this emphasis on the psychological over the biological is potentially liberatory. Heredity, genetics, and the hormones of the female body are not considered to determine behavioral patterns. Rather, an attempt is made to draw a distinction between sexual instinct as a specific aim localized in the excitation of the reproductive and genital apparatus (the earlier biological model) and sexual instinct as represented in the mind as drive (the psychoanalytic model).

Like other criminological discourses of the 1930s, this study is concerned with using the abnormal or perverse behavior to create a model of normativity. What is important here is to consider how the transgressive woman becomes symbolic of all women. Anyone, under the right conditions, may gravitate toward the perverse (and homicidal!) behavior exemplified by Sada. The authors follow Freud in regarding "sexual perversities" to have developed out of a normal constitution but the analysts make little specific reference to Freud. Sada's behavior is found to illustrate not a psychopathic *vita sexualis* but a deviation from normal human behavior produced by an immature sexuality with limited access to a normal sexual object. The authors seek to prove the general thesis that sexual perversions constitute underlying components of normal human sexuality which emerge under particular circumstances and mark a regression to an earlier libidinal phase: "We have reached the conclusion that Abe Sada's psychology is undoubtedly abnormal but not pathological and not unrelated to the psychology of the ordinary person. In her case we discern a caricature, an exaggeration, an amplification or primitive form of the human, especially the female, sexual psychology."[14]

While the authors of the *Psychoanalytic Diagnosis* are stylistically diverse and the focus of their essays is distinct from each other, this study as a whole expressed two intertwining points regarding Sada's deviancy; both tell us much about what the experts thought caused a woman to develop into a "bad girl." The bad girl is hysteric and infantile, yet none of this is of her own doing.[15] The first emphasis regards repression. Sada is described as unable to resist satisfying her instinct. Interestingly, Takahashi argues that because she is unable to repress, she is a hysterical personality. She primarily exhibited a hysterical personality, he argues, as a result of her strong ego, lack of restraint, and propensity to take immediate action.[16] This is an unusual interpretation of hysteria that Freud found to occur within the mechanism of repression. Furthermore, in standard psychoanalytic thought, hysteria is found primarily in the phallic and oral libidinal spheres,[17] and the contributors question whether Sada even entered the phallic stage, in part because of her interest in sexual acts beyond copulation.

The second main point in the study is that Sada is found to have regressed to the infantile stages of pre-Oedipalized sexuality. Perversions Sada exhibited are considered by the authors to express the pre-genital organization of the subject. Precocity is also included as proof of her undeveloped self. The authors find that Sada has not managed to pass through the excruciating test of infantile sexuality. She has not graduated to accept the phallocentric sexual norm. Sada is both unrepressed (operating at the level of instinct) and undeveloped (exhibiting a regression to earlier libinal stage).

In *The Psychoanalytic Diagnosis of Abe Sada* it is the sexual woman who is the primitive. She is not a hypersexualized racialized body in the way that the savage man is in Freud's model, but she is similarly depicted as voracious, unable to maintain sexual decorum and unable to keep desires in check. This psychological infantilism is said to represent an earlier stage in humanity. Abe Sada is the ontogenetic form of the uncivilized past of human civilization. She is unrepressed (instinctually immature) and exhibits regression (ontogenetically immature). Therefore, while there is a concerted effort to move beyond a somatic understanding of Sada's crime, psychologist Ōtsuki Kenji ends up replicating arguments made in evolutionary theoretical arguments about the progression of species and the civilization of humans. Animalistic sexual drive is interpreted as the impetus to Sada's criminal behavior:

> A lustful person is a simple person but not an evil person. Just as an animal is guiltless, she may also be guiltless. What she did was animalistic. Or rather, insect-like. She is more animal than human, more organism than animal, more insect than organism. . . . the unconscious rooted in a psyche transmitted from animals. Furthermore, the deeply embedded experience of primitive times or infant years is embedded in the deep layer of the psyche. Osada's act was an unconscious act and insect-like because . . . insects act similarly. As many know, [female] scorpions and like insects often eat the heads of the male after copulation.[18]

In this entomologizing analysis of the mating ritual, Sada is described metaphorically as closer to an insect or animal. She is shown to experience not an erotic but an animal sexual drive. This attitude is echoed by Takahashi in the same volume; he refers to instinct and copulation in the animal world to explain her desire. Sada's sexuality is primitive. It is prefigured as primary, ontologically discrete, prior to language and culture. The use of insect and animal examples may be the authors' attempt to simplify the content for the general reader. However, since this is only one of numerous publications on psychoanalysis by the Tokyo Psychoanalysis Research Institute it is unclear whether the general reader was of concern. Abe Sada is primitive, hysterical, animalistic in this bizarre report issued by members of the sexological community.

The scientific and legal interest in Abe Sada waned in the postwar period. After the war, Sada becomes a celebrity of sorts. She is interviewed, stages her own autobiographical play, and is the heroine in poetry and novels. The ensuing section illustrate the strange shift in Sada narratives in which she is no longer the deviant bad girl but a laudatory rebel.

The Abe Sada Incident: A Romance for Postwar Japan

Postwar texts about Abe Sada do not treat her as an intriguing example of perverse female desire in the manner that prewar texts do. Quite the contrary, her desire is found to be normal; she motivates a new type of masculine identification which does not anxiously reject female desire but rather embraces and idealizes it. The still sexualized criminal woman in these texts no longer embodies that which is uncivilized and antithetical to social progress. Rather she is an idealized woman for having achieved sincerity of the flesh, unencumbered by social orthodoxy or sense of propriety. Postwar politics transform Sada into a rebellious figure who is idealized for having been a sincere woman (and therefore) a good girl.

One of many postwar stories about Sada features her as a one-time advice-columnist of sorts in the context of a long-running serial novel *True Story: Abe Sada*—the supra title later changed to *Impassioned Woman of Love* (Jōen ichidai onna])—serialized in the pulp magazine *Conjugal Life* (Fūfu seikatsu) from September 1950 to August 1951.[19] The author Nagata Mikihiko actually knew Sada; she was a member in his theatre troupe where she performed a one-act play, *Shōwa's Amorous Woman* (Shōwa ichidai onna). In his fictionalized biography of "Abe Sada," Nagata urges female readers to give up their bourgeois attitudes toward love. The inclusion of texts purportedly written by Sada interrupts the fetishistic gaze of the narrator in this biography. The novel begins with Sada not as an object of the story—confessing woman, patient, criminal on trial, prostitute—but as reader and critic. "Nagata" who is now a character within his biography has sent "Abe Sada" a copy of Ihara Saikaku's well known fictional work *The Amorous Woman* (Kōshoku ichidai onna, 1684), about the travails of prostitute Oharu. Sada "writes" to the author of the story "Nagata" with her thoughts on this gift from "Nagata." Praising Saikaku's portrayal of a woman's sexual life her letter reads:

> The genius of that novel pierced me to the bone. Unlike today's vulgar erotic novels, that kind of novel is strikingly moving. It stayed with me for three or four days. . . . You're responsible for making me read that kind of book when, in the style of Sagano's "Kōshoku-an," I was cleansing my soul in Izu's "Fukiya" while listening to the sound of the wind whistling through the pines from morning till night.[20]

"Sada" makes her own poetic allusions to the similarity of her situation and that of the protagonist in Saikaku's novel. Just as the readers of *Conjugal Life* might similarly sympathize with Sada's story while rejecting any identification with her, "Sada" empathizes and is moved by Oharu's plight while drawing a distinction between herself and the protagonist. The narrative does not, as prewar Sada stories do, rely on "revealing" the neuroses of the criminal woman.

Nagata's novel described Sada no longer as a criminal but rather as a liberated woman with normal desires. The depiction of Sada has changed dramatically in this postwar text. The "Carnal Adrift" chapter of the novel is introduced with a description of the narrative framework employed: "From here [the story] is written as autobiography." And the "autobiographical" story that Sada tells seeks to normalize sexual play and even so-called perverse sexual behavior. "Sada" states matter-of-factly that she only appeared to be as lascivious as Saikaku's Oharu because most women are not honest about their sexual activities:

> I don't think that I'm that different from other women. Most of the women I've known don't confess the truth because they're liars. I still don't know what they mean when they say that I was perverted until about twenty-four or twenty-five. After thirty, I may have done some frightening things with Ishida, but aren't they really part of a normal married lifestyle?[21]

Here "Sada" accuses women of avoiding their desire, and of treating normal sexual activity as "perverted." References to sadomasochistic sexual play as a normal part of a couple's existence couched in the framework of the ambiguous term "conjugal life" runs throughout the novel with a dramatic conclusion addressed directly to the *Conjugal Life* reader. Speaking in a "letter" to author Nagata, "Sada" admonishes women for being too restrained and circumspect in their relationships:

> The relationship between a man and a woman is not something you can read about in a book. It's not that simple. . . . Perhaps I should call it a merging of the spirit and the flesh, though it's absolutely impossible to explain it in words. I know that I won't meet that kind of person again so you won't see me putting on iron sandals to go out searching for him; but in the one in a million chance that a man like Ishida were in this world, I would gladly throw myself into the inferno, risking even becoming a female devil or being turned into a witch. Attempts to delve into the spirit and the flesh by ordinary women, even in relationships as a married couple, have been too timid. I may sound presumptuous but it seems to me that women don't savor even half the joys of the body. Men and women can be in love enough to want to die together. The life most worth living lies in the arduously researched life of passion, and the discovery of a pleasure two or three times the intensity of what you've felt until now. People often laugh that what I did was too extreme. But for the sake of love, for the love between a couple, and for building a harmonious society, isn't it necessary to grapple earnestly with problems of love and desire? With that, Prof. Nagata and readers, I bid you good-bye.[22]

"Sada" is given the last word in this fictional biography which acts as exhortation to couples to dedicate themselves to the pursuit of sexual pleasure. Women are encouraged to recognize and express verbally their sexual desire as a normal part of life; the reward for a woman's free expression of her desire is nothing less than a "harmonious society."

This is only one of the many postwar instances in which Sada acts as a voice of authority on romance. Sakaguchi Ango was one of many men to write about Sada in the early postwar years.[23] He even interviewed her and followed the published transcript of the meeting with an essay praising Sada as a "tender, warm figure of salvation for future generations" for her sincere passion.[24] For him she embodied the possibility of a new "morality" that would be based in the logic of the flesh.

Beyond her tremendous popularity in the immediate postwar context of liberation and existentialism, Sada inspired films in the 1970s and in the 1990s, a best-selling novel which later was made into a film. Most interestingly, in these new versions, Sada is neither a criminal nor an advisor, nor even the sole star of the show. The focus shifts to her partner, Kichizō, concentrating on how Sada's bad girl sexual play transformed him. In Tanaka Noboru's film *The True Story of Abe Sada* (1975), Ōshima Nagisa's film *In the Realm of the Senses* (1976), and Watanabe Jun'ichi's novel *A Lost Paradise*[25] the contentious subject that emerges out of these works is not only the heroically transgressive woman but also a masochistic male obsessively dependent upon the transgressive woman for his countervalent political position. Her capacity for disruption is recuperated and strategically engaged to proffer a new mode of masculinity—one that is masochistic and (therefore) critical of the status quo because it involves complete capitulation to female desire. Her image works in the service of producing a new counter discourse of masculinity through which totalitarian politics and patriarchal cultural values are critiqued. The premise behind this fantasy of female deviancy is that the male who worships such a subversive figure is himself rejecting usual patriarchal relations. Through the identification with female deviancy, the male accomplice can deny any complicity with power and privilege. The female criminal is useful because she can be successfully deployed in romanticizing the rejection of cultural and ideological encumbrances. In this context it is not surprising that Sada's crime of passion re-emerged as a popular touchstone for exploring this new postwar subjectivity and its concomitant celebration of the body and the flesh as a response to the undesirability of prewar paternal power and privilege. Below, a more detailed look at each of these texts illustrates the way in which the masochistic lover of Sada embodies a critical perspective on paternal privilege.

Ōshima's film *In the Realm of the Senses* joins numerous 1970s texts based on the Abe Sada incident in which the male masochist serves a similar allegorizing function. In them, male masochism is cultivated as a rejection of political and military power.[26] It was likely Sekine Hiroshi's poem "Abe Sada" that sparked a comparison between the private devotion of Kichizō to Sada and the public devotion shown by imperial soldiers during the 2/26 Incident, which was an attempted coup d'etat by radical pro-emperor military men who murdered a number of officials and attempted assassination of others in order to seize control of the government in the name of the emperor.[27] Sekine's poem begins with a reference to this attempted military coup which occurred three months before Kichizō's murder. The first-person voice (of "Sada") introduces the quiet epic poem: "The 2/26 Incident / gave rise to war / the year of my Incident / gave rise to a love not yet over."[28] This contrast between public military oppression and the private freedom of the boudoir portrayed in *Realm* is portrayed one year earlier in Tanaka's *The True Story of Abe Sada*. In an early scene in *True Story*, Sada watches military soldiers march by and closes the shutters. When a maid tries to open them, Sada forbids her saying, "The outside light disturbs us." Ōshima takes this contrast between the private desire and public duty further in a poignant scene in which Kichizō in *yukata* and sandals saunters to an inn in a direction opposite a stream of soldiers in starched suits parading before cheering crowds waving the Japanese imperial flag. *Realm* more dramatically affirms the pursuit of masochistic pleasure and complete withdrawal from society as a laudable response to a country on the brink of war. The masochistic lover Kichizō is politicized as an anti-authoritarian rebel for his rejection of male privilege and devotion to the domineering mistress of the boudoir. The spectacle of male masochistic pleasure enables an implicit critique of normative codes in the 1930s when the incident took place. Through the figure of the masochistic lover who willingly sacrifices himself to the fancies of his "Sada-istic" lover outside the institution of marriage and reproduction in wartime, *Realm* celebrates the erotic consuming passion of a couple who is interpreted as implicitly rejecting the bourgeois system that ensures its survival through sexual repression and familial ideology. Kichizō in particular, as the hero of Ōshima's film, represents a challenge to the rigorous social order for his rejection of male privilege and selfless devotion in serving a woman's every desire. "Perversion" is privileged to critique a military economy that makes no room for pleasure.

In rewriting the violence of Sada's act and turning her sadistic lovemaking into a story of romantic love, *Realm* denies the woman the position of "top." Kichizō is understood to be allowing Sada to expel paternal law. Portraying Kichizō as enabling Sada's murder of him undoes the terrifying power that the sadistic woman can represent. Furthermore the question remains as to whether Sada can be identified with undoing the status quo since she still resides in the position of unempowered woman earning her keep through one of the only lines of work available to her at this point in her life, sex work. Nevertheless, the romanticization of their antisocial devotion to each other is at the core of both films. In Tanaka's film, the devotion of Kichizō to his lover Sada is illustrated in his dress. Throughout the film, he wears only Sada's red undergarment. Even in outdoor scenes, Kichizō wears only her red robe. He has so withdrawn from society that he needs only the lingerie she provides him. The two exist in a quiet world that becomes an irresistible web for Kichizō, symbolized grotesquely in a scene in which Kichizō, lonely for Sada who has gone out to retrieve money from a patron, pulls apart a clump of Sada's combed out strands of hair from a tissue, stretches it out like a spider's web, and licks it.

Unlike the original account in which Sada is portrayed as a jealous lover, in this film Kichizō is the jealous one. He is jealous of any time that Sada spends away from him and while he is clearly worried over her angry eruption when he suggests going home to his wife, he gives himself over to Sada completely, seemingly aware that she has plans to murder him and remains loyal to her desire. The film centers on the obsessive love of two people who give recognition to no one except each other. This is especially displayed in a comical scene in which an irritated geisha plays the *shamisen* while Sada and Kichizō, almost completely naked, share food.

In Tanaka's film, *True Story*, Sada's "perversion" is contextualized through a series of flashbacks and voice-overs. After "Sada" has committed the murder, she is depicted walking a literal and metaphorical lonely road while a voice-over recounts her childhood and young adult life. The film flashes back to Sada as a child watching *tatami* mats being made at her family's business, followed with another flashback of her father angrily pinning her kimono sleeves to a large *tatami* mat. The voice-over by "Sada" explains that after she lost her virginity at 15 she became "a geisha, a restaurant maid, an entertainer, a bar girl, a mistress, a prostitute, café waitress, and an escort. There wasn't a day I didn't spend with a man." Through the voice-over "Sada" speaks of wandering from place to place— from Yokohama, Toyama, and Shinshu to Osaka, Nagoya, Tanba no Sasayama and Kobe as we see another flashback of her running away from pimps on railway tracks. The source of her perversion seems to be answered by these references to her first being forced into sex work, followed by her treatment by underworld men. Her hopeless environment drove her to attach herself to Kichizō and her wish to prolong that happiness led to murder. *True Story* raises the same question that we see in the serial novel of whether she didn't do what any woman in her position would do through a song that plays as "Sada" is shown evading the police:

> I did it because I loved him too much / I loved the man's thing / If you think about it, I was stupid / Aah! I did what anyone would do / I did it because I loved him so much / I loved Sada and Kichizō, the two of us / If you think about it, I was stupid / I did what no one would do . . . If you think about, I was stupid / I did what anyone would do. . . . If you think about it, I was stupid / I did what no one would do.

The murder, highly romanticized in this melancholic theme song, is explained as being due to a desperate situation and loving a man too much. *Realm* also replicates this romanticism in its conclusion, which ends with the camera focused on the couple zooming out seemingly through the ceiling of their inn to frame Kichizō and Sada together on the floor facing upward.

A more contemporary story based on the Abe Sada incident appeared in the 1990s. Originally serialized in the *Nikkei shimbun*, Watanabe Jun'ichi's *A Lost Paradise*, also treats the Abe Sada incident as a love story and relies upon the incident to describe an intensely sexual romance between two fictional characters named Kūki and Rinko.[29] The novel follows a man and woman who fall into a perverse and sometimes violent love affair, and who leave their spouses and their jobs to eventually commit love suicide. The couple in this drama is depicted as nothing short of heroic for cutting themselves off from the routine of work and home and eventually life itself. Essential to the symbolism of this romantic tale of social self-ostracization was the well-known story of Sada and Kichizō's romance. The love between Rinko and Kūki is portrayed as utterly synonymous with Sada and Kichizō. The first mention of the Abe Sada incident occurs during the couple's adulterous tryst at Lake Chuzenji:

> Rinko turned toward him nestling close. "My God, that was incredible."
>   For Kūki it had been no less so. "I almost died," he said. "Just a little more. . . ."
>   She nodded. "Now you know what I mean, when I say I'm scared."

Yes, she did sometimes say that. So this is what she meant. He retraced the contours of the experience in his mind, then thought of something else. "Kichizō said the same thing."

"Who?"

"The man strangled by Sada Abe." Images of the two lovers floated through his mind, based on what he'd read.

Rinko's interest was caught. Her voice still sounding exhausted, she repeated, "Sada Abe? The woman who did that grotesque thing?"

"It wasn't especially grotesque."

"She mutilated her lover and then killed him, didn't she?"

She knew only the sensational aspects of the case. Kūki, having read up on it in some detail, saw it rather as a deeply human incident between a man and a woman hopelessly in love. "She was misrepresented in the media." He nudged away the lantern, and added quietly in the half-light, "She did cut it off, but that was after she strangled him, not before."

"A woman strangling a man?"

"She'd done it often enough before, during sex. Like you, just now."

Rinko immediately shook her head and clung to his chest. "I only did it because I love you, I love you so much I can't stand it, I almost hate you. . . ."[30]

As the couple's relationship deepens, Rinko becomes more sadistic in lovemaking while her lover Kūki submits himself increasingly to her wishes. The couple withdraws further and further from normal society. Kūki relinquishes his job and leaves his wife and Rinko leaves her husband. Feeling that this world has little to offer, they determine to leave it. A large portion of this popular novel's readership was businessmen who read it on the train to and from work. Certainly, it can be read as a male fantasy of romantic submission based on the Sada incident that depicts a man who rejects social position, prestige, and wealth as deeply loved and heroic. The "bad boy" needs the bad girl's encouragement to buck the system.

From this small sampling it is evident that postwar texts about the Abe Sada incident differ radically from prewar texts. In the postwar texts, the incident is used to depict antisocial behavior as laudable, romanticized subversion of the patriarchal status quo via the figure of the sadistic woman and the masochistic man. "Perversion" in these texts enables a critical position. The radical act of murder by a woman becomes a surprising source for portraying withdrawal from society as a laudable response to militaristic (Ōshima and Tanaka) and bureaucratic (Watanabe) modern life. This kind of portrayal makes the murder of a man by his lover much easier to take culturally because she is considered to have been driven by love. In the case of Kūki and Rinko or Kichizō and Sada, the deaths are imagined as driven by the woman's intense devotion to her lover with the result that the male is treated as a potent actor rather than as a victim.

In the postwar period, the perverse, sexual suspect of the prewar period is replaced with attention to her masochistic lover who is framed as a rebel in his own right for giving up power and privilege to succumb to the wishes of his sadistic lover. The postwar bad girl Abe Sada is not bad simply because she is destructive, seductive, and even murderous. She is bad because she incites antisocial behavior in man. She becomes heroine to the bad boy wanna-be.

Notes

1. Hon no Mori Henshū-bu, ed. *Abe Sada Jiken chōsho zenbun* (Kosumikku Intōnashonaru, 1997), 48–49.
2. Ibid., 10.
3. Sakaguchi Ango, "Abe Sada-san no inshō." *Zadan* 1, no. 1 (December 1947): 36.
4. For further discussion of the poison woman in Japanese literary history in English, see Christine Marran " 'Poison Woman' Takahashi Oden and the Spectacle of Female Deviance in Early Meiji," *U.S.-Japan Women's Journal English Supplement* 9 (1995): 93–110; Mark Silver, "The Lies and Connivances of an Evil Woman: Early Meiji Realism and The Tale of Takahashi Oden the She-Devil," *Harvard Journal of Asiatic Studies* 63, no. 1 (2003): 5–67.
5. For a discussion of sexology in Japan, see Sabine Frühstück, *Colonizing Sex: Sexology and Social Control in Modern Japan* (Berkeley: University of California Press, 2003), and Ryūichi Narita, "The Overflourishing of Sexuality in 1920s Japan," in *Gender and Japanese History*, vol. 1, ed. Haruko Wakita et al. (Osaka: Osaka University Press, 1999), 345–370.
6. Maesaka Toshiyuki, *Abe Sada shuki* (Tokyo: Chūkōbunko, 1998), 164. Sada in testimony expresses that she felt a sense of futility and believed at some points that becoming a geisha was the only path left for her after being raped by a college student when she was young.
7. Awazu Kiyoshi et al., eds., *Abe Sada: Shōwa jū-ichi nen no onna* (Tokyo: Tabatake Shoten, 1976), 35.
8. See Fukushima Akira et al., *Nihon no seishin kantei* (Misuzu Shobō, 1973), 31–60 for the full report.
9. See Horinouchi Masakazu, *Abe Sada shōden* (Tokyo: Jōhō Senta Shuppan Kyoku, 1998), 224–225.
10. William Johnston, *Geisha, Harlot, Strangler, Star: A Woman, Sex, and Morality in Modern Japan* (New York: Columbia University Press, 2004), 139.
11. Ōtsuki Kenji, *Abe Sada seishin bunsekiteki shindan* published by Tokyo Seishin Bunseki-gaku Kenkyūjo. The main author is listed as the "renowned psychologist Ōtsuki Kenji." A similar pathographical publication on the Sada incident was purportedly published in *Seishin bunseki* (September/October, 1936).
12. Johnston,*Geishan, Harlot, Stranger, Star,* 157.
13. Ōtsuki, *Abe Sada seishin bunsekiteki shindan*, 1.
14. Ibid., 2.
15. It should be noted that the authors hold sometimes conflicting points of view though the individual authors are clearly attempting to establish a fixed hermeneutics.
16. Ōtsuki, *Abe Sada seishinbunseki no shindan*, 38.
17. Ōtsuki Kenji, "Josei aiyoku mondai toshite no Osada jiken," in *Abe Sada no seishin bunseki no shindan*, 68–69.
18. Ōtsuki, *Abe Sada no seishin bunseki no shindan*, 46.
19. Nagata Mikihiko, "Abe Osada" in *Fūfu seikatsu* (September 1950).
20. Ibid., 214–215.
21. Ibid., 220.
22. Nagata, "Abe Osada" (August 1951): 192–193.
23. Sakaguchi, "Abe Sada-san no inshō," 36–38.
24. Interview in "Abe Sada-Sakaguchi Ango taidan," *Zadan* 1, no. 1 (December 1947): 30–35.

25. *Jitsuroku Abe Sada* (The true story of Abe Sada), and *Ai no koriida* (In the realm of the senses); Watanabe Jun'ichi, *A Lost Paradise* (translated by Juliet Winters Carpenter, Tokyo: Kodansha International, 2000).
26. There is a substantial cultural history of masochism and empire in postwar Japan. In literary terms, these historical and political situations translated into a new complex literary or performative trope of the masochistic male who embraces his weakened position.
27. Sekine Hiroshi, *Sekine Hiroshi shi-shū: Abe Sada* (Tokyo: Doyōbijutsusha, 1971).
28. Ibid., 13.
29. Watanabe, *A Lost Paradise.*
30. Ibid., 183.

私たちは

本当の意味でパートナーになったの

バチェラーパーティー❤END

Watanabe Yayoi, final page of "Bachelor Party" (*Kaama*, vol. 7, p. 70), © Yayoi Watanabe/My Way Publishing/*Kaama* 2000.

6

Bad Girls Like to Watch: Writing and Reading Ladies' Comics

Gretchen I. Jones

Introduction

A woman bound with leather, a gag her in mouth, is violently and repeatedly raped by her stepbrother and his buddies. Next, an older man armed with handcuffs and vibrators assaults her. Frame after frame features graphic scenes of similar forced penetration with a variety of implements. Yet at the end of the story, the woman falls into the arms of her stepbrother, declaring he is the only one who knows how to give her pleasure, and she wants to be his sex slave forever. It hardly seems possible that this type of material would be intended for a female consumer, since it seems to fit stereotypical ideas of male fantasies, particularly with the emphasis on forceful penetration and group rape. Yet in Japan, the notorious genre of pornographic "ladies' comics," known for its graphic sexual descriptions and images that frequently depict women being sexually hurt, beaten, or humiliated, is created *for* women, *by* women—and sells hundreds of thousands of copies every month.

On the surface, ladies' comics stories appear to reinforce the "Madame Butterfly" image of Japanese women as subservient, long-suffering, and passive. It might be easy to label the female creators of these texts "bad" in that they seem to promote stereotypical ideas of women by depicting characters taking masochistic pleasure in having men inflict pain and humiliation on them. At any rate, it would be a stretch to state that ladies' comics present a view of sexuality and gender relations in which partners respect one another and are equals. By the same token, it might be easy to declare that female consumers of these texts are "bad girls" for seeking out and apparently enjoying images of women that are far from liberated.

This chapter explores the implications of women creating and looking at these pictures of masochistic imagery. In particular, I am concerned with what it means for women in Japan to *like*, and indeed *choose*, to create or consume images that portray the abuse, rape, and humiliation of women. What does it mean when

Japanese women create these texts, and when female consumers purchase them in such great numbers? What is the appeal of these images, and what is to be gained from creating and consuming them? Can we locate a form of "agency" within these images of passivity, pain, and subordination? One feminist critic has written that "Masochism presents a theoretical challenge to feminist politics because of its apparent relish of subservience in place of an abhorrence of oppression."[1] Linda Williams expands on this notion, stating "it seems imperative for feminist scrutiny to 'open up' to the feminine pleasures of fear."[2] I propose, therefore to accept the "theoretical challenge" offered by the texts under consideration and see what happens when we "open up" and try to understand the pleasures these texts may contain. Below, I introduce the origins of ladies' comics, and then turn to a close look at specific stories typical of these magazines. After considering their production side, I focus on what we know about the readership for ladies' comics, and conclude with some thoughts about how ladies' comics might be interpreted.

What are "Ladies' Comics"?

The term, "ladies' comics," in Japanese *rediisu komikku*, refers to a specific genre of comics aimed at adult, heterosexual female readers. The genre originated in the 1980s out of *shōjo* (young girl) manga that has been immensely popular in Japan since the 1950s. However, as the readership of *shōjo* manga matured, so did their interests. Readers and manga creators both wanted content that matched their new, adult concerns: marriage, family, and sex. The first comics that referred to themselves as "ladies' comics" emerged in 1980, and the genre took off from there.[3]

As the content became more sexually explicit, the sales grew in number and by 1991, when the genre was at its peak, there were over 50 titles with an estimated publication of over 100 million copies per month. Although the genre has seen a decrease since the high point of the early 1990s, ladies' comics still sell well and show no indications of fading. However, protests from various civic and governmental groups and even court cases regarding sexually explicit depictions in the media (not targeted at ladies' comics per se) caused large, well-respected, mainstream publishers to curtail the sexual content in their publications in general, leaving smaller firms with limited catalogues and less public presence to publish material that might come under attack. Consequently, although large publishers such as Kodansha and Shūeisha each presently publish several ladies' comics titles, these comics contain little sexually explicit material. On the other hand, publishers such as Sun Publishing now focus nearly exclusively on sexually graphic material, and have several sexually graphic ladies' comics titles on their publication roster. This chapter concerns the second type of ladies' comics, which are nearly wholly sexual in their content. Regardless of the publisher, however, sexually explicit ladies' comics are widely available, even in corner convenience stores and bookshops. Many can also be ordered online.

The drawings in ladies' comics give one the impression that they must be created by, and for, men, based on their depiction of female characters that appears to differ little, on the surface, from male oriented pornography such as Japanese *ero-manga* or even some images in *Penthouse* or *Hustler*. The female body is drawn in such as way as to appear "on display" to a reader/viewer.[4] Prominence is given to showing the breasts (which are often very large). Moreover, women in ladies' comics drawings are frequently depicted as being raped, or otherwise forced into sex. The vagina and anus are often the focus point of drawings, but in an unusual way: while it is evident what is being shown, the actual area of the vagina and genitals (even on men) is obscured, usually by a blurring or "white-out," or, on occasion, by flowers, a swirling pattern, or other object. However, one of the most interesting aspects of ladies' comics is that primarily women create the drawings and stories. In at least one instance, the magazine editor is a woman as well, and most ladies' comics also have women on the editorial staff. There are several male creators who write under female-sounding pseudonyms (although all ladies' comics creators write under pen names), and some ladies' comics seem to involve men in the creation of content more than others. Nonetheless, by and large, the content for ladies' comics is generated by women.

Surprising as it may seem, ladies' comics readership is also largely female. Several features of the comics show that they are targeted at women. First, and foremost, are the story lines, which concern, in nearly every instance, female protagonists and follow fairly typical romance formulae. Content in comics aimed at men concerns vastly different subjects (baseball, motorcycles, etc.) and features different narrative lines. For example, men's comics would likely not push the point of finding an ideal mate with whom to settle down, but a substantial percentage of ladies' comics stories end with scenes of marriage. Second, the advertisements in the magazines are clearly female oriented: ads for diet products, and hair removal and breast enlargement; want ads for telephone sex and other types of sex business workers; loan service promotions and, more recently, advertisements for videos and sex toys, including vibrators for women. Although there is some anecdotal evidence that men do occasionally read ladies' comics, these advertisements suggest that the product sponsors believe they are targeting a female readership. Third is the emphasis on reader input. All ladies' comics solicit reader feedback and even story ideas through surveys which, when returned to the magazine, may win a prize. There are also advertisements for new (and even established) authors to send in their drawings and story ideas. Finally, most ladies' comics also feature short reader comments in the margins of the stories. Readers write about their sexual experiences, particular acts they like or dislike, or their concerns over their sexual relationships. I will discuss these reader comments in greater detail further. The incorporation of reader comments, always from a female point of view and concerning matters pertaining to women, further supports the idea that these comics are read mainly by women.

A "Bad Girl" Creator: Watanabe Yayoi

Often called the Queen of Ladies' Comics, Watanabe Yayoi is one of the most prolific creators in this genre. A graduate of Waseda University, Watanabe began her career as a *manga-ka* (comic artist) when she was nineteen, publishing in the *shōjo* manga *Hana to Yume* (Flowers and Dreams). She quickly moved on to ladies' comics, where she now receives top billing and is featured nearly every month in at least one (if not more) ladies' comics title. Watanabe's stories are nearly always filled with images of violence toward women; in particular, she portrays female characters who appear to relish physical and psychological abuse, humiliation, and even rape. In some cases, the protagonist is well aware of her masochistic tendencies, and the story shows how she acts on those desires. Other stories are more ambivalent, depicting women who are terribly mistreated, but who come to realize at the end that they like such treatment.

Watanabe often takes up characters whose sexual life is not fulfilling; through masochistic experiences, they find greater pleasure. For example, "Bachelor Party" (*Kaama*, July 2002), opens with a woman named Machi stating, "Since I am marrying a man who is sexually boring, I'm really going to let loose on my last night as a single woman."[5] We learn that Machi has been experimenting with sadomasochism during one-night stands. Her fiancé, Satoru (which means "to realize"—an interesting choice of names, given how the story ends), is an "an elite graduate of T. University and has a nice personality. It's just that sex with him is . . ." according to Machi. Because she is sexually unfulfilled with Satoru, Machi uses her cell phone to solicit partners who like "soft s/m," but she refuses to see any man more than once, asserting that she doesn't like to "sow the seeds of trouble." After a prenuptial dinner with Satoru, Machi spends her last night as a single woman by again soliciting a sexual encounter, this time with several partners at once. She receives a call on her cell phone, and arrives at the designated hotel room where she greets the three nude men who await her. Machi is stripped, bound with leather and blindfolded. After some time (and several pages of various sexual acts with the three men), the "leading man" of the evening arrives. It is clear that he is a novice at this kind of sexual game. On hearing his voice, Machi guesses that the new man is none other than Satoru himself. Soon, her blindfold falls off and her identity is revealed to Satoru as well. The next scene shows the couple in Western wedding wear, walking down the aisle. The following scene depicts their first night as a married couple, with a bountiful selection of leather s/m goods in the background. The image after that shows Satoru binding Machi in black leather straps. The final frame shows Satoru and Machi, body to body against a backdrop of flowers with Machi wearing the same black leather bonds, and the caption "We have become, in the true sense, partners . . . ."[6]

Until the end of the story, the significance of the title "Bachelor Party" is unclear. Given that the opening page states that Machi intends to "let it all hang out" the night before her wedding, the impression throughout the story is that the "bachelor" party must really mean a "bachelorette" party for Machi. Thus, the prenuptial prerogative normally accorded to men—the bachelor party—is co-opted by the woman, at least at the beginning of the story. At the end, however, the meaning of the title "Bachelor Party" finally becomes perfectly clear: Satoru indulges in his own bachelor party the night before the wedding, at which he happens to run into his future bride. It is at this "bachelor party" for Satoru that the couple discovers their shared sexual tastes and realizes (*satoru* in Japanese) that they are truly meant for one another.

Watanabe's story line in "Bachelor Party" (with the s/m one-night stands as well as sex with Satoru) allows for multiple sexual encounters of both s/m and "vanilla" types. Of the thirty-page story, only four pages do not depict sexual acts. Despite this emphasis on the physical and sexual, the story ends with a declaration of true love. The implication is that finding one's perfect match is not only about physical appearance and a successful career, but also about finding a sexual match as well. Moreover, it may take time (for both women and men) to fully recognize one's desires and sexual proclivities, but wedded bliss awaits those who do (in fact, the name Machi is also a homophone for the verb "to wait").

Despite a fairly positive message, the language and imagery used is, true to Watanabe's style, shocking. Machi is called a bitch dog by her various partners, and images of forceful penetration abound. But at each turn, it is made clear that Machi both seeks out this type of sexual encounter, and gets pleasure from it.

The few scenes that show her "everyday" life (outside of sex)—from her elegant wedding dress to her Western style bedroom with a plump double bed, floral sheets, and lots of pillows—reveal that she lives a life of comparative ease, but finds it pales in comparison with the pleasure of her violent, often demeaning, sexual lifestyle. In this regard, Jane Juffer's book, *At Home with Pornography*, is instructive in pointing out that women use pornography to "reconcile the world of fantasy with their everyday lives."[7] Thus, while Machi's sexual fantasies and encounters might fall outside typical norms, the glimpses of her exceedingly average everyday life might also reassure readers that "normal" women may have fantasies of this type.

Although "Bachelor Party" features a woman who seems to know what she wants and has no qualms about getting it, another Watanabe story, "The End" (*Manon*, July 2003), offers a more ambivalent picture.[8] "The End" has a considerably more developed plot than "Bachelor Party," and is not unlike a Harlequin romance with multiple characters and a complex plot. The opening page states, "It was my older brother who made me a woman," leading readers into a familiar motif in ladies' comics: incest. Michi and her brother, Shun, have different mothers, but both mothers are dead. Over the years, Michi has come to depend on her brother for many things, not the least of which is frequent sex. Although Shun says he cares for Michi, his inner thoughts reveal his hatred for her. Through the course of the narrative, we learn through graphic images that Shun had witnessed his own mother having sex with another man, and much later in the story, we learn that his mother, a "harlot," had actually sold him to a gay man as a boy toy (again, a visual image accompanies textual explanation) and later committed a double-suicide with her lover. These diversions from the plot allow for the depiction of a range of sexual acts, including one of two men together.

After the death of his mother, Shun is brought back into his father's house, where a new wife, the mother of Michi, comes into the picture. Shun begins a program of terrorizing the family, which starts with attacking and raping the new mother, a traditional, and very elegant young woman. Eventually, the rapes are revealed and the stepmother, in her humiliation over being violated, throws herself in front of a train. Next, Shun focuses on his half-sister, who is now a young woman, and has sex with her. However, Michi comes to see him as her savior, and enjoys Shun's advances. Since Shun's plan is to destroy the family, he connives to have the father witness sex between his two children. The father loses his sanity after this and soon dies; a scene follows that shows two children making love in front of his funeral memorial. Shun, now head of his dead father's company, next arranges for Michi to marry an old, and exceedingly ugly company employee—who also has sadistic proclivities. Soon, Michi divorces him and wanders the streets, where she is raped by a gang of hoodlums. Although Shun has been intent on destroying everyone, including Michi, once she is actually out of his control and a missing person, he realizes that he "could not even consider a life without her. . . . When you lose something, you find out just how important it is." Months later, Shun locates Michi in a hospital, addicted to drugs, suffering from an STD, and obsessed with sex. Shun declares he'll pleasure her anyway she likes, and the story ends with the line, "Lovers forever. . . . The line separating love and hate is paper thin."

Several key concepts can be extracted from this story. First, is the disturbing notion—common in much men's pornography and in ladies' comics as well—that rape and brutal sexual encounters can actually bring out a woman's sexuality. The first line of the story, "It was my older brother who made me a woman," under-scores this point. Indeed, Watanabe uses this idea quite often in her story lines. At the same time, however, Watanabe's concluding statement that the "line separating love and hate is paper thin" is a reflection of the paradox of masochism: that there can be pleasure in pain. This might lead a reader to think more deeply about the ambivalent nature of desire itself. However, this story can also be read as a tale of revenge by a son against his father, with women merely serving as the tools of revenge. The scene showing Michi and Shun having sex in front of their father's altar at his funeral is telling: although the father is dead, he still has his eye on his children, looking out at them from the memorial photograph. Indeed, it is the father's gaze that motivates much of the plot. First, it is the father's own gaze—seeing his daughter and son having sex—that leads to his death. Simultaneously, that same gaze brings the son the revenge he seeks, by causing the daughter's insanity. Ultimately, however, it is all of these gazes that bring on the realization of pleasure for Michi, and for readers as well, because it is the various gazes that create the numerous graphic images of sexual scenes to which we, as readers, are privy.

"Bad" Girls Who Like to Look: The Readers

Ladies' comics readers are a notoriously difficult group to pin down, and I have never had a woman admit to me that she reads ladies' comics. Yet, the publication runs for each title of ladies' comics are in hundreds of thousands per month; with between 20–40 different titles currently in publication, that comes to a sizeable number of women presumably buying and reading ladies' comics.

What is known about readers comes mainly from the comics themselves, for ladies' comics incorporate readers' ideas and comments in a number of substantive ways. For example, nearly all sexually explicit ladies' comics include a reader survey in every issue, offering prizes (which range from jewelry to dildos) for completing and returning it. The surveys ask readers to rank each story and feature in the issue, and to describe their reading habits (what other ladies' comics titles they buy, to provide ideas for special articles and the like). Additionally, readers are asked to respond to questions about their sexuality: for example, if they have had telephone sex, or masturbated in public, and to describe their experiences. In their responses, readers include detailed sexual fantasies, narratives of unusual sexual encounters, suggestions for new story lines, and requests for particular imagery. Manga creators often place great stock in these reader comments and try to incorporate them into their work as much as possible.[9] In part, this explains how the graphic sexuality and violence in ladies' comics has escalated to its current level: as an editor of ladies' comics notes, "For better or for worse, female comic creators are very serious about their work. Once the popularity for s/m themes emerged, the creators bought erotic books and started learning how to draw the stories submitted by readers with painstaking detail."[10]

Ladies' comics editors also pay attention to these reader comments. *Kaama* was particularly active in soliciting reader participation.[11] The magazine incorporated reader comments and questions from surveys into its "Readers Free Space," where the editor declared, "if it is constructive, any comment is welcomed. *Kaama* is based on our readers' input."[12] In addition, *Kaama* maintained a 24-hour hotline, and printed some of the conversations between the hotline operator (one of the three people on the editorial staff) and the callers. The July 2000 issue also featured three stories with photographs of readers engaging in sexual activity. One, titled "Make me your slave, just for today!" featured a reader whose "dream experience" was to engage in s/m and bondage. The narrative described how Kiyoko (a pseudonym), a 29-year-old married housewife, had been interested in s/m since she had picked up a copy of the "Story of O" in a used book store while traveling abroad. The three-page story documents, with text and photographs, the various acts to which Kiyoko was subjected. At the end of the sexual session, Kiyoko left saying "I had a great experience today. Today, my dream of masochistic sex has really been satiated."[13]

Reader comments and letters are featured in special sections and regular monthly columns, including reader surveys about sex, reader "confessions," and reader profiles. Many ladies' comics titles also include short comments and questions from readers in the margins of the comic stories themselves. Although the authenticity of these comments is difficult to assess, some of the material is without a doubt reader-supplied, based on readers surveys I have seen. That said, whether the sexual experiences women write about were actually experienced, or products of their imaginations, is another matter all together.[14]

I'd like to consider this final type of reader input in more detail. The readers are generally identified by their age, occupation, and place of residence. One reader writes, "It was with my husband during sex . . . and we were doing 'it' in the dark . . . when, suddenly he slipped it into my anus, and easily at that! How was it, you ask? Actually, I had been wanting to experience anal sex. But my husband said, 'I can't deal with that kind of perverted stuff.' I wonder, will I just have to find another partner?"—Hyogo Prefecture, housewife, 24 years old. *Komikku Amour's* editorial staff writes back, "Can't handle that kind of perverted stuff, he says? Your husband's an unusual type these days. And, even if he says it went in by mistake, he DID slip it right in!"[15] This comment is typical in that it takes up sexual activity outside of procreational, heterosexual sex and demonstrates an interest in that type of sex. Perhaps most telling, however, is the suggestion that either in "real" life or in her fantasies, this reader is grappling with her own erotic desires, and is by no means passive in her gratifying her sexuality.

Another reader writes, "A long time ago (when I was 17), when I was having sex with my boyfriend and we had a lot to drink, we had 'outdoor s/m!' I was tied up with a rope to a cherry tree. My feet barely touched the ground, and then along came two other men, and it was three men to just me. There was no way to resist, but it felt good, and we went crazy doing it again and again until dawn. My boyfriend was really turned on. It felt so good that it was almost scary."—Saitama, company employee, 29 years old. The editor writes back, "Being strung up to a cherry tree is so literary, but who were these men that appeared, and where was this???"[16]

One function that these reader comments perform is that the authors/creators of these comics and the readers seem to acknowledge one another, and work together to create content that "works." The reader-supplied comments in particular seem to suggest to other readers that there are other women who consume these comics, and have similar problems or concerns. This form of "sisterhood" is similar to an observation Linda Williams makes in her re-consideration of Hitchcock's *Psycho* and its female viewership. She asserts that women enjoy being in the company of other women and being scared together, forming a connection between women.[17] The reader comments in ladies' comics may function in much the same way: creating connections between female readers. At the same time, male aggression can be transformed into a game that women can safely manipulate, secure in their position as the sole arbiters of their fantasy world.

What seems clear from reading these reader comments is that those women who write in communicate a genuine eagerness to experience a range of sexual experiences. While some have various problems, such as concern over postnatal weight gain, extramarital affairs, and lack of sexual interest in their husbands, many women are also interested (whether in reality or just in their fantasies) in experimenting sexually, and feel little shame or compunction over doing so. Indeed, the tone of these women readers strikes me as showing a significant degree of comfort with their sexuality. Through the forum of reader comments, readers are accorded a mode through which they can express themselves; it offers an additional titillating bonus of being a public—albeit anonymous—forum, by appearing in print. The agency one might assign to the production of erotica then, is clearly displayed, and available to all those who send in their comments.

Interpreting Ladies' Comics

Ladies' comics readers have been much maligned, particularly in Japan, for consuming these comics and the female sexuality they portray. For example, Fujimoto Yukari, a vocal commentator, believes that Japanese women are unable to express their sexuality overtly, and sees a link between the violence in ladies' comics and readers' difficulties in "owning" their own sexual desire. Fujimoto asserts that taboos on female sexual expression in Japan often make it difficult for Japanese women to psychologically "let go"; a rape fantasy makes it possible to circumvent these social proscriptions without incurring guilt for doing so. According to Fujimoto, depictions of gang rape, also common in ladies' comics, further minimize any active responsibility on women's part and function as a convenient plot device to prolong and increase the number of sexual scenes depicted. Ladies' comics are thus seen as an outlet for inherent but never openly acknowledged sexual desire.[18]

Sakamoto Mimei, despite being a ladies' comics author herself, is even more condescending of the sexuality expressed in ladies' comics in her 1997 article, "Yappari onna wa baka datta" (Women were fools after all).[19] Far from publishing stories that express women's independence and autonomy, Sakamoto argues that ladies' comics contain stories in which responsibility for sexuality still rests solely with the man, and she is critical of conservative "Cinderella" stories in which the heroine is saved by a "prince" who marries her in the end. Sakamoto thus faults ladies' comics authors and the women who purchase them for not taking their sexual desire into their own hands.

As my earlier discussion of reader-supplied comments indicates however, ladies' comics readers are far from passive, and many readers seem quite open about acknowledging, expressing, and perhaps even acting on their sexual desires, even masochistic ones. While it is true that depictions of gang rape are common in ladies' comics, Fujimoto's assertion that a gang rape fantasy reduces responsibility seems mislaid, although such depictions do, as Fujimoto indicates, expand opportunities for various dramatizations of sexual activity. Sakamoto's critique of ladies' comics focuses more on the content of the individual stories, which often admittedly follow a "Cinderella model."

Indeed, Fujimoto and Sakamoto are joined by a long line of critics who find fault with female-centered genres, including "low" cultural forms such as Harlequin romances, horror films, soap operas, and nineteenth-century sentimental literature. Even female critics tend to deride genres typically associated with women. As Tania Modleski writes in her groundbreaking study of soap operas and Harlequin romances (what she terms "mass-produced fantasies for women"), "Women's criticism of popular feminine narrative has generally adopted one of three attitudes: dismissiveness; hostility—tending unfortunately to be aimed at the consumers of the narratives; or, most frequently, a flippant kind of mockery."[20] Fujimoto and Sakamoto (both female critics) seem to take a dismissive and at times hostile stance toward ladies' comics and their readers, as do most academics in Japan with whom I have discussed the genre.

However, as Modleski and others have shown in their work, such "mass-produced fantasies for women" are not the simple, stereotypical narratives that one might think. Modleski and Radway show that there is a complex relationship between the viewer/reader and the text. Women are not, in their analyses, passive consumers but actively engaged participants in their consumption of popular texts. As discussed earlier, in ladies' comics, the reader's voice is heard in a variety of ways; the interactive nature of ladies' comics is one of the genre's distinguishing features.

For example, Erino Miya, billed as a "ladies' comics critic" by the magazine *Takarajima*, states that "because ladies' comics emerged as a new genre, readers and manga creators together explored what sorts of comics women would respond to, and thus the genre could develop in tandem with the changes in women's consciousness." Erino points out that in the late 1980s and early 1990s, Japanese women saw a wide range of social changes, including those brought about by the Equal Employment Opportunity Law (EEOL). In contrast to Sakamoto, Erino asserts that with the passage of that law "the reality of putting all one's dreams into love and marriage was lost. What appeared after that was a new dream—sex." As Erino puts it, "Instead of the fantasy of happiness through love and marriage, why not pursue the fantasy of 'a changed self through sexual orgasm'?"[21]

By focusing on the interaction of readers and creators, and concentrating on how women's ideas and knowledge of sexuality have developed, Erino assigns a more active role to readers—one that consciously acknowledges the "fantasy" inherent in *both* the ideal of marriage *and* the ideal of fully autonomous sexual desire. Erino thus contrasts what she views as an active acknowledgment of fantasies of desire on the part of ladies' comics readers with the passive attitude seen by Fujimoto and Sakamoto. A sanguine approach to sex and sexual desire certainly exists among some ladies' comics readers, as this reader's comment reveals:

> Actually, it's fairly easy for a woman to pick up a ladies' comic and read it. When women are interested in sex, it's not like we can just go rent an erotic video like men do, and when we can't stand it any longer, unlike men, we can't go to a soapland [the Japanese equivalent of a massage parlor]. So, even though women's interest in sex may be high, the reality is that there are few means for us to choose to enjoy sex. Although many ladies' comics are rather crude, women acknowledge that and buy them because it's better than nothing, and when they are finished reading, they just toss them in the trash.[22]

Inquiries into pornography and its consumption also indicate that reading or viewing these kinds of texts is not a passive activity. As Jennifer Wicke writes, "When the pornographic text or image is acquired, the work of pornographic consumption has just begun." By "work," Wicke means "the shuffling and collating and transcription of images or works so that they have affectivity within one's own fantasy universe—an act of accommodation, as it were. This will often entail wholesale elimination of elements of the representation, or changing salient features within it."[23] Thus, although a ladies' comics story may show a male taking the initiative in sexual acts (as Sakamoto claims) or images of passive women, the *act* of consuming these images itself is by no means passive. The reader of a ladies' comic may skip scenes or entire stories, read faster or slower, or focus on particular scenes or images, depending on how each element or story affects her "own fantasy universe." The reader responses I surveyed back this up: some readers like certain images and scenes, while others reject them; the same applies to readers' reactions to particular ladies' comics authors and their narratives or drawings.

The "work" Wicke describes of "reshuffling" images and ideas to fit one's personal requirements also indicates that a story or image does not necessarily have a universal meaning for each consumer. Though one might expect that a woman reading a ladies' comic would identify with the heroine, Judith Butler has argued that fantasy "does not entail an identification with a single position within a fantasy; the identification is distributed among various elements in the scene."[24] Elizabeth Cowie notes that whereas in reality gender roles are relatively inflexible, in fantasy gender boundaries can be transgressed with impunity.[25] Thus a reader can play with assuming or identifying with a number of different subjectivities within the fantasy regardless of his or her gender or sexual orientation, and might enjoy this flexibility.

The nature of fantasy also permits a reader to "safely" manipulate and experiment with other dichotomies, such as those of active and passive, subordinate and dominant. In fact, it is not uncommon to find in a single ladies' comics issue one story that features a heroine who is brutally abused and another in which a female character is a cruel dominatrix who forces men to perform humiliating acts for her entertainment. Of course, fantasies of rape, violence, and domination are not limited to ladies' comics; pornographic materials of lesbian, gay, and even heterosexual men frequently involve fantasies of both dominance and submission.

In fact, what we are confronting is the manifold nature of desire itself. Elizabeth Cowie's view on this issue is illuminating. She writes that desire "is most truly itself when it is most 'other' to social norms, when it transgresses the limits and exceeds the 'proper.'" She goes on to characterize desire as "the transgression of the barriers of disgust—in which the dirty and execrable in our bodily functions becomes a focus of sexual desire."[26] Ladies' comics clearly transgress a range of "boundaries"— surely part of what makes them at once fascinating and abhorrent to so many.

Transgression of boundaries, and the complex processes involved in reading ladies' comics texts indicate that their readers do indeed possess a form of agency. It might even be that ladies' comics readers have taken a further step, as Ellen Willis suggests about women who consume pornography:

> A woman who is raped is a victim, a woman who enjoys pornography (even if that means a rape fantasy) is in a sense a rebel, insisting on an aspect of her sexuality that has been defined as a male preserve. Insofar as pornography glorifies male supremacy and sexual alienation, it is deeply reactionary. But in rejecting sexual repression and hypocrisy—which have inflicted even more damage on women than on men—it expresses a radical impulse.[27]

In part, Willis' statement explains how ladies' comics may appear "deeply reactionary," for many of the images and scenes portrayed in them do seem to support gender stereotypes, and the frequent representations of violence toward women may indeed be indicative of the reality of women's social position. At the same time, such representations undoubtedly serve a purpose of making language and ideas available, thus opening the door to some form of agency. As Nicola Pitchford points out in a statement that echoes Willis, imagery of a pornographic nature "offers readers the possibility of shifting, contextual, and multiple relations to power." Nonetheless, Pitchford also recognizes that these same relations are "always vexed, never freely fluid."[28] Ladies comics, then, illustrate both normative gender relations as well as a site for possible transformation.

Moreover, Jane Juffer's point that it is essential to understand exactly how women access pornography is relevant to the case of ladies' comics as well. Since women are largely responsible for the creation of content, and, women are able to purchase ladies' comics with little difficulty, women in Japan do not have to go out of their way to access this particular brand of pornography. This, according to Juffer, is an essential component in assigning agency.

Ladies' comics challenge us to rethink and possibly to reassess our ideas about women, pornography, and desire. Although the images and ideas contained in ladies' comics may seem incompatible with the aims of feminism, as Marianne Noble asserts, "failing to recognize the powers and pleasures women have sought in exploiting dominant discourses of desire will impede the transformation of desire."[29] We need to understand how women appropriate these discourses to discover how they both serve and subvert the status quo.

Indeed, journalist Kiriyama Hideki sees ladies' comics as an agent of change. He observes that despite the increasing diversification of women's lifestyles, so that they are no longer tied down by the "career or marriage" choice, at present ladies' comics, with their readers' contributions, are one of the few vehicles that provide a voice for women's deepest desires. The Japanese media at large, Kiriyama argues, still does not nurture the true (*honne*) voice of women.[30] Questions of whether there is one "true" voice for all women aside, the very existence of ladies' comics and their tremendous sales indicate that their readers are demanding access to avenues of sexual expression that have been solely male until now. And, in fact, a more positive and proactive view of female sexuality has become evident in ladies' comics over the years. A March 1999 *Komikku Amour* advertisement for videos for women says, "For you women who have been having sex for the sake of men, it is time to graduate. Now is the time for you to discover your own body and desire—sex that will satisfy you!"

Notes

I thank Kiyomi Kutsuzawa, editor of *U.S.-Japan Women's Journal*, for granting kind permission to include here some portions of my article, "Ladies' Comics': Japan's Not-so-Underground Market in Pornography for Women," *U.S.-Japan Women's Journal English Supplement* 22 (2002): 3–31.

1. Robin Ferrell, "The Pleasures of the Slave," in *Between Psyche and Social: Psychoanalytic Social Theory*, eds. Kelly Oliver and Steve Edwin (Lanham, MD: Rowman and Littlefield Publishers, Inc, 2002), 19.
2. Linda Williams, "When Women Look: A Sequel," in *Senses of Cinema: An Online Journal Devoted to the Serious and Eclectic Discussion of Cinema* (SoC) 2001. Online at <http//www.sensesofcinema. com/contents/01/15/horror_women.html>.
3. For a more detailed discussion of the ladies' comics genre and typical contents, see my article " 'Ladies' Comics': Japan's Not-so-Underground Market in Pornography for Women," *U.S.-Japan Women's Journal, English Supplement* 22 (2002): 3–31.
4. See Deborah Shamoon, "Office Sluts and Rebel Flowers: The Pleasures of Japanese Pornographic Comics for Women," in *Porn Studies*, ed. Linda Williams (Durham, NC: Duke University Press, 2004), 77–103, for more on the depiction of the female body in ladies comics and parallels with visual conventions in *shōjo manga*.
5. Watanabe Yayoi, "Bachelor Party," *Kaama* vol. 7 (June 2000): 39–70.
6. Ibid., 70.
7. Jane Juffer, *At Home with Pornography: Women, Sex, and Everyday Life* (New York: New York University Press, 1998), 20.
8. Watanabe Yayoi, "The End," *Manon* 10, no. 7 (July 2003): 4–103: The so-called *yaoi* comics, which nearly exclusively feature sex between men, are also erotic comics targeted at women. For more on *yaoi*, see Akiko Mizoguchi, "Male-male Romance by and for Women in Japan: A History and the Subgenres of *Yaoi* Fictions," *U.S.-Japan Women's Journal English Supplement* 25 (2003): 49–75; and Kazumi Nagaike, "Perverse Sexualities, Perverse Desires: Representations of Female Fantasies and *Yaoi Manga* as Pornography Directed at Women," *U.S.-Japan Women's Journal English Supplement* 25 (2003): 76–103.

9. Personal interview with Miya Chie, May 1993. Sakamoto Mimei, a manga creator herself, discusses reader responses to her own work; see Sakamoto Mimei, "Yappari onna wa baka datta." *Shinchō 45*, 16: 7 (July 1997): 234. Erino Miya also makes reference to reader comments, as does Karasawa Shun'ichi; see Erino Miya, "Redikomi no orugasumu ga josei no sei ishiki o kaeta?" *Bessatsu Takarajima 30* (August 1994): 130–133, and Karasawa Shun'ichi, "Redikomi no taiken tōkōsha wa, naze kaiinu to yatta hanashi made kokuhaku shitaji no?" *Bessatsu Takarjima* 240 (1995): 192–202.

10. From an interview with a male editor, as cited by Erino, "Redikomi no orugasumu ga josei no sei ishiki o kaeta?," 132.

11. *Kaama* ceased publication in the fall of 2002.

12. *Kaama* 7 (June 2000): 294.

13. Ibid., 282.

14. A manga creator graciously supplied me with approximately fifty handwritten reader surveys dating from 1992–1993. The handwriting differed on each survey, as did the content, leading me to believe they were submitted by actual readers. By and large, the letter writers stated that they wanted more "hard-core" content, and were candid in describing their sexual lives.

15. *Komikku Amour* 14, no. 8 (164) (August 2003): 28.

16. Ibid., 40. The image of a woman tied to a cherry tree is indeed "literary," and is most likely the figure of Yukihime, a character in Kabuki and the subject of a number of ukiyo-e prints.

17. Linda Williams, "When Women Look: A Sequel."

18. Fujimoto Yukari, "Onna no yokubō no katachi Rediisu komikku ni mieu onna no seigensō," in *Nyū feminizumu rebyū 3: Pornogurafii: Yureru shisen no seijigaku*, Shifuji Kayako, ed. (Tokyo: Gakuyō Shobō, 1992), 73–74.

19. Sakamoto, "Yappari onna wa baka datta," 238–239. The "boom" in Ladies' comics has often been linked to the bubble economy, and the subsequent decline in sales linked to economic decline. See, e.g., Kiriyama Hideki, "Kanojo ga 'kagekina manga' o yomu riyū," *Purejidento* (September 1993): 192.

20. Tania Modleski, *Loving with a Vengeance: Mass-Produced Fantasies for Women* (Hamden, CT: Archon Books, 1982), 14.

21. Erino, "Redikomi no orugasumu ga josei no sei ishiki o kaeta," 133.

22. Kiriyama, "Kanojo ga 'kagekina manga' o yomu riyū," 195.

23. Jennifer Wicke, "Through the Glass Darkly," in *Dirty Looks: Women, Pornography, Power*, ed. Pamela Church Gibson with Roma Gibson (London: British Film Institute, 1993), 70.

24. Judith Butler, "The Force of Fantasy: Feminism, Mapplethorpe, and Discursive Excess," in *Feminism and Pornography*, ed. Drucilla Cornell (Oxford and New York: Oxford University Press, 2000), 491.

25. Elizabeth Cowie, "Pornography and Fantasy: Psychoanalytic Perspectives," in *Sex Exposed: Sexuality and the Pornography Debate*, ed. Lynne Segal and Mary McIntosh (London: Virago Press, 1992), 141.

26. Ibid., 134.

27. Ellen Willis, "Feminism, Moralism and Pornography," in Ellen Willis, *Beginning to See the Light: Pieces of a Decade* (New York: Knopf, 1981), 223.

28. Nicola Pitchford, "Reading Feminism's Pornography Conflict: Implications for Postmodern Reading Strategies," in *Sex Positives? The Cultural Politics of Dissident Sexualities*, ed. Thomas Foster, Carol Siegel, and Ellen E. Berry (New York: New York University Press, 1997), 21.

29. Marianne Noble, *The Masochistic Pleasures of Sentimental Literature* (Princeton, NJ: Princeton University Press, 2000), 8. See also Janice Radway, *Reading the Romance: Women, Patriarchy, and Popular Culture* (Chapel Hill: University of North Carolina Press, 1984).

30. Kiriyama, "Kanojo ga 'kagekina manga' o yomu riyū," 195.

Book cover, *Shopping Queen* by Nakamura Usagi (Tokyo: Bunshun Bunko, 2001). Reprinted with permission from Bungei Shunjū.

7

Branded: Bad Girls Go Shopping

Jan Bardsley and Hiroko Hirakawa

With her party imminent, Muffy was as hyperactive as a Japanese tourist in a Louis Vuitton outlet store.[1]

Plum Sykes, *Bergdorf Blondes*

Introduction

The Japanese fascination with high-end luxury goods has captured worldwide attention for good reason. Japanese have become by far the largest group of consumers of American and European designer-brand fashion goods, their purchases accounting for 40 percent of this multi-billion-dollar global market.[2] In 2002, a Merrill Lynch analyst stated that the dozen largest fashion houses in Europe owed one out of every three dollars in worldwide sales to Japanese consumers.[3] Websites for leading brands such as Cartier, Chanel, and Louis Vuitton include Japanese-language pages and many high-end boutiques abroad regularly employ Japanese-speaking sales personnel.[4] Among these brands, Louis Vuitton has been particularly successful in Japan since it was first promoted in 1977 by a new magazine for young women, *JJ*: By 1999, an estimated 40 percent of Japanese women owned at least one Louis Vuitton product[5] and more than half of Japanese women in their twenties had a Louis Vuitton handbag.[6] When Louis Vuitton opened its flagship store in the stylish Omotesandō shopping district in Tokyo in September 2002, it earned $1 million in opening day sales.[7] Store openings for Hermès and Prada in 2001 were remarkably successful as well.[8]

The visibility of this consumption, especially among young, single women, provoked reaction of all kinds in Japan in the 1980s. Celebrations of conspicuous consumption took many forms, including foreign shopping tours, gold-dusted desserts, and a best-selling novel (*Somehow Crystal*, 1981) that also functioned as a shopping guide.[9] The "bubbly life," as popular writer Hayashi Mariko dubbed it, was initially explained by the booming Japanese bubble economy of the 1980s and the increase in single women who wanted to delay marriage, live rent-free with their parents, and spend the money gained from their low-level office jobs on leisure activities such as shopping, travel, and entertainment.[10] In 2005, long after the bursting of the bubble economy, the age of first marriage is still rising, the birth rate still falling, and luxury goods are still selling extraordinarily well in Japan. The success of Western designer brands cannot be explained by the tastes of single working women alone. Teenagers and matrons alike are part of the market, too—or, as some would say, equally part of the problem.

Brand love has also given rise to moral panic, cautionary tales, and stories of heroically thrifty mothers. In 1989, for example, the climactic final year of the bubble economy, "A Bowl of Buckwheat Noodles," a short story by an anonymous male writer about a young, frugal widow raising her two young boys on her own, became a hit, and was later made into a film.[11] Conservative (male) politicians immediately saw the value of this story as a counter-narrative to tales of spoiled young women in Japan, and praised this sentimental story as a much-needed gem. When asked in 1985 what he was looking for in a wife, Crown Prince Naruhito emphasized that he wanted a partner who shied away from luxury and that he'd be troubled by someone who became excited over "this and that at Tiffany & Co"[12] (although in 1993, the prince presented his bride with an enormous diamond ring). More recently, a 2003 ladies' comic warned readers against excessive spending by running a reportedly true story describing how a young mother's purchase of a pricey handbag, bought to keep up with richer, brand-conscious neighbors, sent her spiraling downward into lying, gambling, and bankruptcy.[13]

Extreme shopping can get one into trouble, but does purchasing a Western designer brand really qualify the consumer for inclusion in this volume devoted to *Bad Girls of Japan*? If so, what about such a purchase stigmatizes it as "bad," and how does this indictment not only serve to disparage the individual consumer but women in general, producing "bad girls"? In turn, how do these definitions of deviant shopping frame opposing notions of consumer correctness? This chapter explores these questions by investigating a range of discourses in Japanese popular culture that purport to explain how women should and should not consume fashion products.[14] Sources include recent fashion guides, a novel, a top-selling series of shopaholic confessional books, feminist debate, and a popular television drama. Considering these sources, all directed primarily to female audiences, allows us to understand what girls' and women's motivations are *imagined* to be and how these imagined desires are deployed to influence brand consumption and encourage feminine virtue in contemporary Japan.

Stories that tie feminine virtue to careful shopping are not new in Japan, though they have changed over time. The long-held ideal of the Japanese woman as the good wife, wise mother who practices frugality, saves religiously, and makes sacrifices for the sake of family and nation, although still part of the Japanese imaginary, seems quaintly old fashioned when compared to the lifestyles often promoted to women today. Self-expression has replaced self-sacrifice as the new orthodoxy and permits a certain degree of self-indulgence as well, especially when it's in service of developing a sophisticated palate. In popular fiction from the 1970s to the present, the figure of what we call the enlightened career woman has embodied this ideal. Her years of hard work have not only enabled her to buy luxuries on her own, but have given her the refined taste to appreciate them without losing her personal sense of chic. A deserving consumer, the enlightened career woman emerges as the good girl of brand shopping, and simultaneously represents an embrace of the new opportunities open for women in post-EEOL corporate Japan.[15] Bad brand-girls come in many forms in contemporary popular culture, the nature of their transgression defined differently by different critics. They are reprimanded as badly behaved, as badly duped, as badly overindulgent, as badly impulsive, and, most often, as badly lacking in a mature sense of self. The loudest retort to such charges comes from our featured bad girl, author Nakamura Usagi, self-described Queen of Shopping and brand addict, who disdains the good girls of shopping and defends the bad as launching a legitimate protest against the constraints forced upon women in Japan. Interestingly, images of the wise wife continue to circulate, but as psychologist Ogura Chikako shows, this new wife's wisdom echoes Nakamura's cynicism: she aims to beat the system, getting her brand fashion and having her self-expression, too.

Elegant and Entitled: The Enlightened Career Woman

The bad girls of brand shopping turn up regularly in fiction and fashion guides devoted to celebrating the deserving brand consumer. Functioning as the foil in these narratives, the badly behaved girls' immaturity highlights the worldliness of the enlightened career woman. They play the annoying hick to her borderless chic. To discuss the politics involved in this hierarchy, we begin by introducing two fictional examples of the enlightened career woman: Kawashima Midori, the heroine of the hit TV drama, "Brand" (2000), and Kitamura Sami, the lead character in Hayashi Mariko's best-selling novel, *Cosmetics* (2002).[16] We also describe how the same hierarchy operates in Suzuki Rumiko's 2001 fashion guide, *Don't Take Hermès Lightly or You'll Be Sorry*. This section concludes with some thoughts about how "Brand," *Cosmetics*, and *Don't Take Hermès Lightly* pave the way for the continued consumption of luxury fashion—as long as one plays by the career woman's new rules and does not sink to becoming a bad girl.

Styled as single, attractive, 30-something urbanites, Kawashima Midori and Kitamura Sami enjoy management positions in the fashion industry and exciting, complicated love lives. They resist parental pressure to marry, are absorbed in their work, earn a good salary, and ultimately triumph over the challenges presented by their competitive work environments. Their consumption of chic fashion reflects and rewards their hard-won career achievements and their sense of personal identity. Although they face disappointments and agonizing decisions, both revel in taking control over their lives—a theme strongly emphasized in the narratives. At the conclusion of their stories, still chic and single, Kawashima Midori and Kitamura Sami are happy with their lives and themselves. In past decades and for some critics today, such concentration on creative work, multiple suitors, and resistance to marriage and motherhood would brand these characters bad girls or at least defiant ones. But "Brand" and *Cosmetics* depict Midori and Sami as bravely struggling to embrace the new orthodoxy of personal expression.

Several aspects of Kawashima Midori's connection to luxury brands are relevant here. As the press manager at Dion Japan, an elegant French boutique (modeled on the Christian Dior firm, which consulted on this production), 35-year-old Midori must plan fashion shows and arrange advertisements. Always elegant, Midori (played by Imai Miki) rarely wears a recognizable brand, although the camera lingers on what appear to be her expensive handbags. She is never shown shopping or even thinking about what to wear. What does consume Midori's attention is her search for the true meaning of her life. A beautiful, white silk Dion blouse, a memento of her late mother, symbolizes authenticity for Midori. Shots of the blouse, which was a romantic, first-anniversary gift from Midori's father, are soft and brilliantly lit so that it appears radiant, pure, and almost other-worldly. Midori learns that her mother did not initially wear the garment, but had to develop a self-confidence that would make her feel as if she were not "giving into the blouse." The stronger the mother's sense of self-worth became, the more comfortably she could wear the blouse. Midori grew up closely associating the fine blouse with her mother, as if it were a part of her. Over the course of the drama, Midori looks at the white blouse, wondering if she, too, will discover her true self and be worthy of the blouse and her mother.

Kawashima Midori is angelic. She venerates the craftsmanship that goes into Dion products, generously forgives the competitors and subordinates who betray her, and works so hard that the company finally forces her to take a vacation. One of Midori's self-defining moments in "Brand" arises when she must confront publicity attacking Dion advertising as leading to teenagers' compensated dating (*enjo kōsai*) practices, in which flirtation and sex with older men is traded for expensive brand fashions. Against all advice from her superiors, Midori declares at a press conference that Dion does indeed bear some responsibility. But her most important message, one that she describes as her own and not Dion's, is that young girls must learn that it is they "who make the brand shine," that a brand's authenticity is nothing without their own. Here, Midori brings in the example of "a friend's" path to self-discovery and how the friend matured into an elegant white blouse.

Midori decides to quit Dion after this incident and also turns down marriage proposals from two handsome Japanese suitors: her former lover, the brilliant marketing manager at Dion and her newer and true love, a man ten years her junior, who is heir to a long line of tea ceremony practitioners. In the drama's final scene, one that blends French haute couture and the quintessentially Japanese, Midori, resplendent in her mother's blouse and a white pant suit, walks buoyantly through a grove of cherry trees abloom with pale pink blossoms. In a voice-over, Midori says that she has realized that authenticity lies in the discovery of her deepest self, and when she discovered that, she says in the final line of the drama, she knew that happily, "I had become my own brand."

Whereas Midori is goodness personified, 30-year-old career woman Kitamura Sami, although presented sympathetically, is frankly ambitious and even cunning. Chief of the press section for a French cosmetics firm in Japan, Sami learns to be wary of the competition. She has an edge that scares men away from long-term relationships. The liaison to which she continually returns is with another executive, a cynical, brilliant, sensuous—and married—Japanese man with a European flair for fashion, who gives Sami advice on how to reach ever greater career heights. Luxury brands figure in Sami's narrative as a means to measure her success. She evaluates her life as if examining it through the lens of a fashion advertisement. When she goes to meet a potential mother-in-law, Sami wears a Chanel suit that she bought on a shopping tour in Singapore with the intention of displaying herself as a successful, self-supporting career woman. At the outset of the novel, Sami sits at a sidewalk café in Paris, wondering whether or not to spend $4,000.00 on a Birkin bag at Hermès. Turning over the reasons to buy the handbag (more color selections and a better price in Paris than in Tokyo), Sami considers how this decision entwines with choosing what she will do next in life, particularly in her career.

Kitamura Sami is no angel, but she is not the badly behaved Japanese girl of Parisian shopping. She and her friend Kanako, who resides in Paris, complain about the "waves" of Japanese tourists pushing and shoving in the Gucci and Prada boutiques, beside themselves to grasp the latest items touted by the magazines back home. Sami makes sure her fashion sense sets her apart from these Japanese, even at De Gaulle airport:

> Sami's suitcase, made by Zero, was fashioned in metallic gold. It was not an ordinary sort of suitcase, the kind that Sami disliked, but one that she had sought out in Hong Kong. It came equipped with casters and handles, and was intended to be pulled upright. There were many women who were hunched over enormous suitcases that they had to push and drag. Sami couldn't stand the horrible clanking noise those made. Sami's philosophy of beauty held that, especially in an airport, a woman must walk smartly. No doubt Sami stood out among all the ordinary young women lining up at the economy counter. In situations like this, Sami would not be caught dead wearing jeans like them. Instead, she wore slim slacks bought in Paris and topped them with an old Agnes B. leather coat.[17]

At the end of the novel, Kitamura Sami is on a plane to Paris for a major fashion event that she has organized. While she remains fashionable and involved in the cosmetics industry, her growing maturity has given her a certain distance from designer brand fashion, which she imagines proves her more sophisticated than other women. Soaring across the world from Tokyo to Paris, Sami knows that "all the choices to make in her life are hers alone."

The hierarchy of the elite career woman as the deserving consumer and the naive one as the bad girl is most overtly expressed in fashion guides like Suzuki Rumiko's book, *Don't Take Hermès Lightly or You'll Be Sorry*.[18] Suzuki argues that the truly self-aware, elegant woman does not like brands simply because they are brands as naive consumers do, but feels a natural affinity with them. Suzuki includes herself among the elite in this homage to boardroom chic:

> Each had much experience in her own line of work and each had her say. They epitomized the label, "Career Woman." When I glanced down at their hands, I saw they were all using the same kind of notebook. It was the same Hermès notebook that

I had. Was this merely coincidence? These women were not at all the sort to pursue trends or mimic others, nor were they friends who enjoyed dressing alike. These women, who had achieved position and worked with such enthusiasm, were naturally at home with a brand appropriate to their status. It was their competence in their work and their maturity, I suspect, that led them to choose quality items that did not follow trends. That is the point of having top-quality things. That is, I believe, what makes one deserving of Hermès.[19]

Don't Take Hermès Lightly is a how-to guide geared to women, which sports a paper cover that resembles Hermès packaging. Deep, dark orange in color, the cover is made to appear pebbled like leather and looks as if it is wrapped in the brand's trademark slender, brown ribbon which here, too, bears the imprint "Hermès—Paris." Purchasing this volume for a mere 1,600 yen (around $15.00) can make one feel as if obtaining an elite Hermès product at a bargain and also as if capturing a bit of the elite career woman's glamour. The book is not so much a guide to Hermès products as a primer on how to become the discerning woman who deserves to own them. The guide does not pretend that modeling such behavior will bring career success, only style.

Enter the bad girl. Suzuki instructs through testimonies to the career women described above but, even more forcefully, with horror stories about spectacularly unqualified Japanese shoppers. As Suzuki sees it, the sad state of Hermès consumption by Japanese has resulted from newly moneyed consumers' lack of self-awareness, their distasteful shopping habits, and their failure to appreciate the fine craftsmanship of Hermès. Much like Sami's lament over Japanese shoppers in Paris, Suzuki describes how shocking French people find it when Japanese crowd into chic boutiques, dressed down in sloppy sneakers and blue jeans as if they were going on hikes, pawing all the goods, yelling excitedly to each other, and behaving in a demanding manner toward the poor sales people forced to serve them. Suzuki recounts the scene of a shop suddenly engulfed by a wave of uniform-clad school-girls on a fieldtrip from Saitama, imagining that the incident must surely have frightened the French into wondering if a Japanese cult had landed in Paris. She advises that if you are so young that you need to buy on credit, have your parents buy your luxuries, or work hours at a part-time job to obtain your designer bag, then you are really too young for Hermès. Young or old, she warns, one must never stoop to buying an Hermès knock-off. Echoing the admiration for authenticity extolled in "Brand," Suzuki holds that the most pathetic of all consumers are those who possess a real Hermès product but are themselves fakes. In contrast, Suzuki's enlightened career woman has developed a personal authenticity that draws her to intrinsically fine things, which simply happen to be designer brands, and has a restraint that meshes with the traditions of the tea ceremony. Suzuki sees the career woman's inner and outer polish combining to form a seamlessly perfect Total Look. This sophistication renders her borderless, a cosmopolitan with global panache, who would never be mistaken for one of the fakes among waves of Japanese pushing and shoving at Gucci.

The enlightened career woman, as embodied by Kawashima Midori and Kitamura Sami, and extolled by Suzuki Rumiko has earned a special admission into the world of high French fashion, almost passing for a European aristocrat. Yet references to the aesthetics of the Japanese tea ceremony and cherry blossoms show that the career woman is no poor step-cousin to European high taste, even as she rejects contemporary mass culture in Japan. As anthropologist Karen Kelsky has discussed, some women in Japan of different ages and classes have used the idea of the West, as an imagined space of utopian possibilities, to launch a critique of Japanese society. Kelsky observes how this process has tended to focus on a project of developing "good manners" that "entails above all a disciplining of the body, mind, and voice in accordance with the 'rules' of a predeterminative West. . . ."[20] In this female competition over achieving class through classiness, the desirability of the luxury fashion itself goes unquestioned. Reading the brand good solely as a sign of personal authenticity shifts scrutiny away from conglomerate profits, global marketing schemes, and a more complex politics of class and gender in Japan.

Confessions of a Dis-Eased Shopping Queen

Some people like to say in the most pretentious way, "Oh, no, I don't love brand goods simply because they are brand goods, but because they are so well made that we can use them for decades." That's a lie! We buy brand goods because they are brand goods.[21]

Nakamura Usagi, The Shopping Queen

This quote offers a taste of the radically different picture of luxury shopping presented in the confessional books written in the late 1990s and early 2000s by our featured bad girl, Nakamura Usagi. This self-styled "queen of shopping" makes no claims to being elite, enlightened or even very elegant. Indeed, it is Nakamura's unabashedly frank, even bawdy approach toward brand-love that has turned this bad girl into a best-selling author.

Fans of Nakamura's writing are familiar with her biography. Born in 1958, Nakamura Usagi (Noriko) spent some post-college years as an OL, got married, and then divorced in the 1980s.[22] She initially earned fame and fortune as a fantasy novel writer in the early 1990s when her story, *Adventures of Gokudō-Kun* was turned into a successful anime.[23] Nakamura squandered her new fortune, spending her income in brand shopping—approximately 500 million yen over five years—while letting her taxes fall into arrears. In 2001, at the age of 43, Nakamura met a young man working at a host club in Tokyo, became infatuated with him, and ended up spending 15 million yen on him in that year alone. This infatuation also piqued anxiety over her appearance, leading Nakamura to undergo a series of expensive cosmetic surgeries. All the while, Nakamura professed disillusionment with heterosexual love, and was married to a gay man from Hong Kong whom she had met at a gay bar in Tokyo.

As a woman in her 40s with a successful writing career, Nakamura earns more money by far than most Japanese women could ever hope to make. One might expect that a woman of her age and position could present herself as the enlightened, Westernized sophisticate, who happens to buy luxury fashions because of their quality, not their brand cache. Yet, as we shall see, Nakamura's books do everything possible to distance her from this ladylike image. Not only does she openly acknowledge that she hungers after brands because they are brands, she loves to use vulgar male speech and works scatological humor into her shopping essays. Unlike her self-restrained, enlightened sisters, Nakamura comes across as desire itself, ready to devour anything she wants, spewing out the "excrement" of vile language and bodily waste, denying any etiquette that would contain her.

Nakamura claims that we all live in a gendered, capitalist world, and we all want to feel superior and gain worldly success in this world, where every woman is treated unequally because of her gender, class, and nationality. Why bother to make up all these excuses to rationalize this desire, Nakamura challenges. If something helps us *feel*, even momentarily, like a winner in this not so fair world, then why not make use of it? In the process of asking such questions and exposing her own vulnerabilities, Nakamura also exposes the arbitrary nature of the gender, class, and national boundaries that are rendered natural and inevitable through the rhetoric of the deserving consumer. Nakamura is a buster of such boundaries. She is a very bad girl—and not apologetic in the least.

Nakamura's satiric criticism of the popular 1992 tract, *Living Pure and Poor* by Nakano Kōji[24] exemplifies her bad girl attitude. According to Nakano, at the height of the Japanese bubble economy, the materialistically driven, or even worse, the nouveau riche became the dominant image of the Japanese abroad. The Japanese were in danger of being seen, as indicated in the quote at the outset of this chapter, as the frenetic fans of luxury brands. Concerned with the unflattering nature of this international image, Nakano traces the works of such famous forefathers of high Japanese aesthetics as poets Saigyō and Bashō, among others, in hopes of showing that Japan has a long tradition of valuing simple, plain living. (Nakano conveniently neglects to consider the extent to which such simplicity was a pose for these privileged men.) The fact that Nakamura never read the book, as she admits, does not stop her from satirizing "the pure and poor" (*seihin*) by coining a new term, "dirty and poor" (*dakuhin*). Nakamura's dirty and poor live "an evil, gaudy, greedy, and stingy life soiled by materialistic desires."[25] Asserting herself as an embodiment of the dirty and poor, Nakamura challenges Nakano Kōji as follows:

> Yes, I, the dirty and poor, LOVE money. I spend it all, when I have an income (or even when I don't), which, of course, makes me poor. In contrast, the "pure and poor" do have money because they don't spend it. They say, "Yes, I live pure and poor," yet, in fact, they have millions of yen in their bank accounts. Isn't this nothing but cheating? . . . They are the enemy of a capitalistic society. Think of it this way. If we stop spending money, what will happen to our industries? If all of us begin to live self-sufficiently, our farmers, fishermen, and store managers will be in trouble. Our salarymen will become unemployed. Okay, let me be honest and say that I don't really care what would happen to them. What about *ME* if all of you become self-sufficient and stop buying my books? How can I make a living? Are you going to take care of me then, Nakano Kōji?![26]

Nakamura's books direct equal sarcasm to narratives of the enlightened consumer as lauded by Suzuki Rumiko who simply happens to choose brand goods.
She sees this claim as an alibi for brand love. Nakamura asserts that all women who
buy brands do so "because they make you feel superior even though you may be
ugly, dumb, or from a poor family."[27] They are "rewards for the winners of a game
that is called capitalism."[28] She explains as follows:

> Brand goods are so expensive, aren't they? Why?! Well, the way I see it, we lose our
> appreciation of them if they aren't. Imagine a Chanel T-shirt that costs only 500 yen.
> At that price, no one could say, "Hey, I wear Chanel!," and feel superior. If you want
> to present yourself as the "rich-lady-in-her-brand-clothes" type, then even a T-shirt
> needs to cost 100,000 yen . . . Even if it costs 250,000 yen or so, people will still
> forgive that if it is a Chanel or Hermès T-shirt. They need to. But if a Daiei
> Supermarket's original T-shirt cost that much, people would either be carried away
> in shock to the emergency room, or start a riot.[29]

Nakamura claims brand fashions are rewards precisely because they are expensive
and not because of any intrinsically high value. They *need* to be expensive in order
to maintain the illusion of high quality that, in turn, supports the claims to sophistication made by their owners. Nakamura seeks to expose the tale of the deserving
consumer as a capitalist story that naturalizes the hierarchy of tastes marked along
class and national lines. She sees all luxury brand owners, including elite career
women, as aspirational consumers, buying expensive brands in hopes of feeling as
if they have climbed the global ladder of success. She argues that when Japanese
buy Western brand goods over inexpensive domestic ones, they are consuming
this illusory feeling, although the rhetoric of "happening to choose brands" conveniently allows them to avoid admitting that the love of brands is motivated by
this wannabe mentality.

Empire of Marriage, the 2004 book that records a dialogue between Ueno
Chizuko, famous feminist and sociology professor at the University of Tokyo, and
Nobuta Sayoko, a well-known feminist psychotherapist, provides an example of
Nakamura's point about how successful professional women resist being seen as
another wannabe. At one point in their conversation, Ueno criticizes the people
lining up at the openings of European brand stores in Tokyo:

> It's so embarrassing to watch them on TV. I can't help but wonder, "In the middle of
> this recession, what's with these women?!" Japan is the easiest market for those
> European brands, isn't it? I, for one, don't own any of those brand goods.[30]

Shyly acknowledging that she, in fact, carries a Louis Vuitton bag, Nobuta explains
that she likes such products because they are sturdy. Adopting the same stance as
Nakamura Usagi, Ueno calls this an excuse, remarking once again how she "takes
pride" in not owning a single European brand and pressing Nobuta to "frankly admit
that people love European brands because everyone knows they are expensive."[31]
In response, Nobuta tries hard to convince Ueno (and herself as well?) that her
admiration of the sturdiness of Louis Vuitton products is genuine, sharing an

episode of how her first Louis Vuitton bag, after having been stolen and found deserted in a river, was returned undamaged half a year later. When Ueno retorts that bags manufactured by Ichizawa Hanpu, a Kyoto firm, are as sturdy as Louis Vuitton's, Nobuta helplessly replies, "Well, but they are not sold in Paris."[32] Ueno pounces on this opportunity to remind Nobuta that marketing the image of Paris is the "global sales strategy for European brands" and that this strategy is apparently working well since Paris is the "added value" that has made Nobuta choose Louis Vuitton over Ichizawa Hanpu.[33] In response, Nobuta insists that she was genuinely moved by the sturdiness of her lost and found Louis Vuitton bag, at which point Ueno exclaims, "That's the reason you came up with afterwards [not the reason why you bought it to begin with], and I can't believe a woman like you would do that!"[34]

Apparently, Ueno, like Nakamura, believes that the educated and sophisticated career woman—like Nobuta—is after all not that different from crowds of other Japanese brand shoppers. For Ueno, these Japanese brand shoppers are easy prey for the strategic sales of European brands, and she takes pride in refusing to become one of these victims. Ueno thus creates, though probably unwittingly, a new category of the deserving consumer and places herself in that category—the consumer who rejects brand shopping because she is sufficiently enlightened to recognize the fallacious rhetoric of intrinsic value. In contrast to Ueno and Nobuta, Nakamura has no problem admitting that she *is* a brand-shopping junkie, and asks, so what? At the same time, Nakamura admits that "the big three French brands [Chanel, Hermès, and Louis Vuitton] that we dumb Japanese women are crazy about" not only place "outrageous prices" on their products, but also do not hesitate to produce "some incredibly useless goods."[35] Just look, she says, at Chanel's 150,000 yen portable umbrella that doesn't even hold up in the rain. Nakamura quickly adds that she bought it anyway because it has a Chanel logo on it.

Certainly, Nakamura uses this satiric self-representation as a "brand-crazed, dumb Japanese woman" to poke fun at the career woman who naively believes that personal authenticity makes her a borderless, brand-deserving cosmopolitan. Yet Nakamura simultaneously distances herself from the intellectual sophisticate like Ueno who virtuously resists brands. She also laughs at pundits, such as one misogynistic TV commentator who accused Japanese women of indirectly supporting French nuclear experiments through brand consumption and castigating them all in effect as bad *girls*:

> And, when I heard of the commentator's accusation, I, a dumb Japanese woman, felt bad. This didn't stop me buying brand goods, though. I am an unpatriotic Japanese citizen, or should I say, an "unpatriotic" global citizen? The "brand-crazed-dumb-woman" is even less honorable than the "*kogyaru*-who-doesn't-hesitate-to-sell-her-underwear-or-even-her-body." Well, these two labels are too long for brand names, so instead, [the commentators] are using "Chanel" as the brand name for us dumb women. Our foreheads are branded with the "Chanel" logo. Say, isn't that cool? . . . Thanks to us dumb Japanese women, the French government must have earned all the foreign currency it needed to engage in another nuclear experiment.[36]

Here Nakamura's mockery obviously targets those who worry over the international image of Japanese women's luxury shopping. Yet Nakamura's cynical joke about being an "unpatriotic global citizen" seems equally aimed at feminist intellectuals who, in an attempt to critique and resist the impacts of global capitalism, sometimes end up being self-righteous. Her sarcasm reminds us that, like it or not, most of us have some wannabe mentality, as it is almost impossible to completely escape from the pervasive forces of late capitalism. To Ueno, who speaks proudly of not owning a single European brand, Nobuta cannot help but say, "Well, I cannot possibly see why we need to be that obstinate about brand shopping."[37] Nakamura would certainly be in amused agreement with this remark.

Nakamura, however, has another reason why she is unable to identify with the puritanical intellectuals who refuse brand shopping. For Nakamura, the pleasure it brings may be illusory, but it is better than having none:

> Brand women are deceiving themselves. We seek to self-actualize. Then, by owning brand goods, we feel we have improved ourselves. "Oh, at last, I have become a woman who deserves to have a Birkin bag." What a self-deception! But it feels so good, and, before we know it, we are trapped in the hell [of endless cycles of brand shopping]. I know this [because I have lived it].[38]

As Nakamura sees it, everyone wants to win the capitalist game, and gender complicates this game. The male winner gains pretty women as his reward. His career success and wealth do not contradict his sense of being a man, and bring satisfaction to his (heterosexual) masculine identity. In contrast, for Nakamura, a successful career does not guarantee a woman male suitors unless she is pretty, and, thus, does not satisfy her (heterosexual) feminine identity. The TV drama, "Brand" produces nothing but "a huge belch" from Nakamura, as she is "sick and tired of being overfed that kind of superficial portrayal of the cool career woman."[39] For Nakamura, the very idea of such self-actualization is a sham. Unlike such nebulous dreams, brand goods are attainable for a price, requiring money and nothing else. But, as Nakamura acknowledges, facing the fact of such displacement is hard to do.

Conclusion: Married, Still Shopping

Associating Japanese luxury shopping with either independent career women or naive teens and 20-somethings assumes that their status as single women plays a crucial role in their self-indulgence. Surely, once they get married and especially when they have children, these consumers will shed their haute ambitions and concentrate on working, spending and *saving* for the good of their families. As feminist psychologist Ogura Chikako argues in her 2003 book, *Conditions for Marriage*, however, many of today's women view marriage as a means for continued self-actualization and pleasure, and not as a gateway to a life of constraint. Their unsentimental views of gender roles, work, and marriage in Japan are ones that ring true for Nakamura Usagi and lend another example of how narratives of self-discovery, the good life, and shopping are intertwined.

In late 1990s, Ogura conducted interviews with single women in their 30s and early 40s, who were living in metropolitan Japan. Interestingly, Ogura found that those interviewees who were graduates of junior colleges or lower-ranked four-year universities had what she called a "new housewife orientation."[40] These women had grown up observing how hard their fathers worked as corporate warriors while leaving their wives to shoulder all responsibility for child-rearing and housework. These wives and mothers also often worked outside the home in part-time jobs to supplement the family income. Witnessing the difficulties of their mothers' double-burdened life led these daughters to consciously reject this path as no longer viable. At the same time, they knew their lack of educational credentials would prevent them from following the path of the elite career woman embodied by Kitamura Sami or Kawashima Midori. Moreover, this option did not appeal to them, for it simply meant that they would live like their fathers, too exhausted to enjoy their private lives. Seeking to avoid the hard lives of their mothers and their fathers, these women are specifically looking for a husband who will allow them to lead a materially comfortable life with plenty of free time to pursue self-actualization. If they find such a husband, they say, they will happily marry and become the stay-at-home housewife. They do plan to go back to work once their children have grown up. But this work should be the kind that will satisfy their desire to find their true self. Stylish work such as the job of an interior designer or a floral shop owner is preferred. For these women, it does not matter if such work can make them much money as long as it brings them pleasure. As they see it, the husband bears all the responsibility for a family's financial support.

As Ogura argues, these women are realistic in their view that without exceptional educational credentials in contemporary Japanese society they can only hope to pursue self and individuality after finding a well-off husband. The new housewife orientation represents a cynical attempt on the part of these women, who are strongly conscious of their marginal status, to get the most out of their condition. It is the desire of these women that renders men as merely a means, rather like brand goods, to realize their dreams of pursuing personal authenticity. By holding onto this desire, they are explicitly refusing to become the conventional housewife, who would put the welfare of the family before her own personal fulfillment. Safely married, with children in tow, they may look reassuringly good, but these women are as bad as any of the single, luxury shoppers. In their case, however, the replaceable, expendable, and consumable item is the husband. As they see it, their real dilemma is to find the rich husband. They are acutely aware that they might have to wait longer to obtain him than a Birkin bag. In fact, waiting years in line for a Birkin bag might be the better bet.

Throughout her books, Nakamura screams, "I'm not pretty. I don't have a rich husband." Thus, despite her successful career, Nakamura voices the dilemma faced by these women, who are marked down as average. She confesses that she feels like yelling at a young wife dressed head-to-toe in Hermès leisurely cruising with her kids at a party: "Damn you!! You have the husband who allows you such luxury, while I have to buy all these myself. Oh yeah, I'm not pretty like you, bitch!"[41] She feels like yelling not out of disdain for the woman's superficiality, but because her presence reminds Nakamura that something is missing in her life. She turns to brand shopping as a way, perhaps the only way, at least to *feel* that she has what she wants. Shopping temporarily satisfies but also frustrates her because she also knows what she really wants is not the brand goods, or not even the rich husband. What drives Nakamura to what she calls "empty consumption" is the persistent "sense of despair that something fundamental is missing . . . but I don't know what that is." This inability to find the missing something, she believes, is due to the way "society cleverly hides this by telling us we already have it."[42] It is a social phenomenon she sees as enabled by the mechanism that forces women to choose between career success and the allure of femininity.

Throughout her books, Nakamura also calls herself, The Naked Empress. She buys brands because she wants to feel like a queen, even though she knows the masses will see through her pretense. The masses will see that she is as vulnerable as any other woman as she struggles to live the contradictions of gender in a late capitalist society. A graduate of a mediocre four-year university, an ex-OL who is divorced, a surgically altered brand shopper, Nakamura is both too cynical and honest to place herself on the side of the enlightened. She is too cynical to believe, like Midori or Sami, in the liberated girl's dream of finding an authentic self through having it all. Nakamura is also too honest to believe that she is free from the binding fetters of established heterosexual gender identity. Nakamura may be a self-proclaimed bad girl, but she is also a wounded one.

Notes

Thanks to Laura Miller, many Bad Girl authors, and the Japanese History and Culture Study Group at Duke University organized by Simon Partner and Gennifer Weisenfeld for advice on this chapter.

1. Plum Sykes, *Bergdorf Blondes* (New York: Hyperion, 2004), 35.
2. Carol Matlack et al., "The Vuitton Money Machine," *Business Week International Editions*, March 22, 2004: 48. Matlack also reports that in 2004 purchases by Japanese accounted for an estimated 55% of Louis Vuitton's $3.8 billion in annual sales and that its executives hope to tap into "rising affluence in China and India".
3. Clay Chandler and Cindy Kano, "Recession Chic," *Fortune*, September 29, 2003: 52.
4. Korean and Chinese language pages are also available at the websites, as of this writing.
5. "Japan's Luxury Product Sales Increase," *IGN Global Marketing Newsletter,* September 1999. Online at <http://www.pangaea.net/IGN/news0051.htm>.
6. Suvendrini Kakuchi, "Despite Hard Times, Shopping Habits Die Hard," *Asia Times Online* January 15, 1999.
7. George Wehrfritz and Kay Itoi, "The Luxury Bubble," *Newsweek*, February 10, 2003: 34.
8. Hermès opened a 12-story, $137 million store in the Ginza on June 27, 2001. According to *Japan Today*, Japanese are "the company's most loyal customers [accounting] for 25% of global turnover for the brand" while "France brings in 17% in sales." "Hermès Opens Glitzy Shop in Ginza," *Japan Today* June 28, 2001. Online at http://www.japantoday.com. Prada's flagship store, which opened in Tokyo's Aoyama district on June 7, 2001, "cost $85 million, Italy's largest-ever investment in [Japan]," Chandler and Kano, "Recession Chic," 52.

9. Tanaka Yasuo, *Nantonaku, kurisutaru* (Tokyo: Shinchō Bunko, 1981). For more on this novel, see Norma Field, "Somehow: The Postmodern as Atmosphere," in *Postmodernism and Japan*, ed. Masao Miyoshi and Harry Harootunian (Durham, NC: Duke University Press, 1989), 169–188.

10. Yamada Toyoko, *Burando no seiki* (Tokyo: Magajin hausu, 2000), 218.

11. The 1992 film, also titled *Ippai no kakesoba* (A bowl of buckwheat noodles), was directed by Nishikawa Katsumi.

12. Kawahara Toshiaki, *Kōtaishi-hi Masako-sama* (Tokyo: Kōdansha, 1993), 104.

13. "Otto ni naisho no shakkin de jiko hasan wa kannō?," *Onna no toraburu*, October 1, 2003: 165–184. We thank Kinko Ito for this source.

14. Space does not permit discussion of the equally interesting discourse in English-language publications on Japanese women's shopping. Analysis of Japanese consumption and domestication of "the West" in broader context is found in Joseph Tobin, ed. *Re-Made in Japan: Everyday Life and Consumer Taste in a Changing Society* (New Haven: Yale University, 1992).

15. Barbara Molony discusses the EEOL in, "Japan's 1986 Equal Employment Opportunity Law and the Changing Discourse on Gender," *Signs* (1995): 269–302; Aki Hirota analyzes celebrations of the new career woman post-1970 in "Image-makers and Victims: The Croissant Syndrome and Yellow Cabs," *U.S.-Japan Women's Journal English Supplement* 19 (2000): 83–121.

16. "Brand," a serialized television drama, aired in 2000. Hayashi Mariko, *Kosumechikku* (Tokyo: Shōgakkan, 2002), 35–36. The novel was originally serialized in the magazine *Domani* from January 1997 to December 1998.

17. Hayashi Mariko, *Kosumechikku*, 35–36. Kaori Shōji reports that over 100 Agnes B. outlets operated throughout Japan in the early 1990s and were so popular that "one out of every three women under 25 owned at least one Agnes B. item." Kaori Shōji, "It's Always Oh, So French in the Ginza," *International Herald Tribune Online* October 13, 2003.

18. Suzuki Rumiko, *Herumesu o amaku miru to itai me ni au*, 3rd edition (Tokyo: Kōdansha, 2002).

19. Ibid., 160.

20. Karen Kelsky, *Women on the Verge: Japanese Women, Western Dreams* (Durham, NC: Duke University Press, 2001), 227.

21. Nakamura Usagi, *Shoppingu no joō* (Tokyo: Bungei Bunko, 2001), 50.

22. Nakamura's given name is Noriko. Usagi, which means rabbit, is her penname.

23. Nakamura Usagi, *Gokudō-kun manyūki* (Tokyo: Kadokawa Bunko, 1991).

24. Nakano Kōji, *Seihin no shisō* (Tokyo: Sōshisha, 1992).

25. Nakamura Usagi, *Datte, hosii n da mon!: Shakkin joō no binbō nikki* (Tokyo: Kadokawa Bunko, 1997), 22.

26. Ibid., 23–24.

27. Nakamura, *Shoppingu no joō*, 50.

28. Ibid.

29. Nakamura Usagi, *Konna watashi de yokattara* (Tokyo: Kadokawa Bunko, 2000), 38.

30. Ueno Chizuko and Nobuta Sayoko, *Kekkon teikoku: Onna no wakare michi* (Tokyo: Kōdansha, 2004), 58–59.

31. Ibid.

32. Ibid., 59.

33. Ibid.

34. Ibid.

35. Nakamura, *Shopping no joō*, 53.

36. Ibid., 53–55.

37. Ueno and Nobuta, *Kekkon teikoku*, 59.
38. Nakamura, *Konna*, 192.
39. Ibid., 227.
40. Ogura Chikako, *Kekkon no jōken* (Tokyo: Asahi Shimbun-sha, 2004), 30–39.
41. Nakamura Usagi, *Pari no toire de s'il vous plait* (Tokyo: Kadokawa Bunko, 1999), 178.
42. Nakamura Usagi, *Shoppingu no joō: Saigo no seisen!?* (Tokyo: Bungei Shunjū, 2004), 243–244.

Girls' *yabapuri* (2005). Used with kind permission of photo takers.

8

Bad Girl Photography

Laura Miller

"Girls tend to hold nothing back, and don't think too much. Without thinking too much, they let their feelings rule their actions."

Araki Nobuyoshi, commenting on the prize-winning photography of the
17-year-old photographer Hiromix [1]

Inside a Tokyo auditorium in 1995, a female high school student beat 456 applicants to become the grand prix winner of Canon's Cosmos of Photography photo contest, earning one million yen in prize money. In the quote above one of the judges is describing his interpretation of the winning entry, entitled "Seventeen Girl Days" by photographer Hiromix (real name Toshikawa Hiromi). Following the contest she embarked on a brilliant career as a successful photographer, publishing photo-diary books with occasional snapshots of herself in underwear or topless. Although the judge intends to praise the putatively unmanipulated nature of girls' photography, Hiromix and other young women do invest thought into their photo diaries and other photographic projects. As the anthropologist David Sapir noted "A photograph is at once a direct representation of reality and the result of an utterly subjective choice." [2] Sapir's insightful acknowledgment of debate over reality and its representation is especially true of the type of photography produced by Japanese girls.

Consider a photograph two girls have produced, yielding something unlike the cute image most people would predict: one girl has drawn thick black eyebrows and enlarged nostrils on her face. Black dots are sprinkled on her cheeks and forehead. The other girl has penned in a pair of sunglasses to cover her eyes, in addition to red clownish circles on her cheeks and a large blue drool descending from her mouth. They each have the character for "stink" inscribed with black marker on their chests, and three icons representing steaming piles of feces are overlaid on all as the finishing touch. It is a photo of "real" girls, but the subjective and thoughtful decisions that contributed to its making are readily apparent. This photograph is striking but not unusual. In its cuteness and its abjection, the photograph is part of a trajectory in girls' photographic culture that dates from the late 1990s and remains vibrant today. It is representative of a new genre of odd, repulsive, or naughty photos that are circulated among girls, plastered on cell phones and notebooks, and collected in thick albums. Although they begin as private media, these photos are also posted on the internet and published in girls' magazines. This chapter examines the phenomenon of bad girl photography, asking what it says about changes in female culture and girls' use of new media. As this phenomenon demonstrates, girls are not simply consuming mass culture forms, but are exercising creative intervention with their own unique modifications. Thriving at the center of contemporary Japanese cultural interest and vigor, girls have been the driving force behind this and many other technological developments. Bad girl photography offers an especially vivid way of understanding the significance of girls' culture as an example of female-centered innovation and swank that is at the heart of cultural trend setting.

Included in my notion of bad girl photography are innocuous or merely wacky photographs that are nevertheless annotated with obscene or offensive words and phrases. Indeed, the majority of girls' photographs are marked up to create "graffiti photos."[3] Why do young women, in a society where the aesthetic of cuteness is so overwhelmingly pervasive, want to take photographs of themselves in which they appear ugly or grotesque? In a nation that proclaims itself home to the most sweetly polite women in the world, why are girls defacing their photos with scatological, obscene, or bizarre writing? Older Japanese dismiss such behavior as part of a generalized decline in civilized conduct among youth, who are usually described as impertinent, self-centered, and lacking in common sense.[4] Yet this is a particularly gendered form of resistance tied to developments that occurred within girls' culture.

Graffiti photos are unregulated cultural production in which anything may become the topic for photo-textual representation, melding an ironic awareness of gender norms with an audacious thumbing of those very ideals. A primary purpose of girls' photography is to strengthen friendship networks by commemorating social groupings and events,[5] and girls' photos frequently contain text that details the nature and circumstances of their ties. The exchange of freakish photos is doubly effective in this regard, netting the recipient in a web of mutual vulnerability. Bad girl photos provide visual-linguistic access to a deep vein of discontent that bubbles beneath the surface of contemporary girls' culture. Clearly aware of how the schoolgirl persona is fetishized, they appropriate or defy such images. Girls' photo art is an index of social change and challenges mainstream femininity norms, something we see in other domains of girls' culture such as fashion, language, and popular music.[6] Japanese gender ideology puts a premium on female self-sacrifice and other-directedness, yet girls are using their photos as expressions of hedonistic pleasure and unfettered experimentation and self-expression. Girls relish giving badness a graphical dimension completely outside the usual venues of adult surveillance. Although Japanese women have often been associated with the domestic sphere, these photos are displayed in easily accessible forums, eroding the distinction between private and public. All these social aspects of bad girl self-photography are the result of female-driven technological developments that stimulate new forms of consumption.

Enabling Technologies

Japanese photographers have been active since the 1850s, their work developing as an important cultural form that changes in tandem with historical and social developments.[7] The infusion of developing photographic technology led to fresh methods for commodifying the female image. Initially, some Japanese were afraid of the camera, and worried that photographs sapped your lifeblood or spirit. A widely circulated rumor was that "If you have a photo taken once, it will dilute your shadow; if you have it taken twice, it will shorten your life span."[8] Japanese and foreign photographers avidly documented famous sites and everyday scenes, including rice planting, funeral processions, and women engaged in activities such as weaving, eating, spinning, and harvesting oysters. Foreign tourists, traders, and diplomats also stimulated a market in brazenly artificial photos of "old" Japan's disappearing culture, which often had to be staged in the studio.[9] These studio photos were delicately hand tinted by artisans, and many of them featured geisha or courtesans. Images of Japanese women were immensely popular abroad, and foreign men went so far as to coerce prostitutes into posing for the camera in order to get these coveted photos. In 1891, Ogawa Kazuma was commissioned to photograph one hundred celebrated geisha for an exhibit, and later he and others published photo books of similar collections. These famous albums enabled anyone to vicariously participate in geisha connoisseurship. Photos of geisha, actors and other appropriate objects of public looking were also printed on postcards and flyers, called "bromides."

Photography achieved the patina of legitimacy when the Empress Meiji and other royals began allowing themselves to be photographed. Prior to the photograph, the idea of having your likeness preserved had been reserved for famous leaders, wrestlers, and entertainers who were immortalized in woodblock prints. As photography became more acceptable, studios offering services to anyone who wanted a commemorative photo proliferated. Photo shop owners also profited by making extra copies of especially pretty female clients, distributing these for sale without their permission, a practice resulting in an 1882 law that a person's image could not be sold without her permission. Interest in women's photographs also led to the first beauty contests.

In 1908, Japanese women competed in a global beauty contest sponsored by the *Chicago Tribune*. Japanese contestants from "good" family backgrounds sent in their photos for newspaper publication, and an international team of judges selected the winners. Events such as this helped transform photos of nonprofessional women into a respectable form, and thereafter women's magazines began inviting readers to likewise submit photos for publication. These photos were accompanied with the girls' names, educational background, and personal characteristics to create a sort of debutante column. Once readers saw photos of "good girls" they rushed to contribute their own pictures, a form of self-presentation that was nevertheless uncomfortably reminiscent of the manner in which brothel workers and geisha were advertised with photo book listings. Even in the first decades following the Pacific War there was ambivalence about this blatant public display on the part of good girls. Those who presented themselves in photographs, especially ones not buffered with text describing family background and educational history, were in danger of being seen as bad girls no different than other public women such as geisha.

By the 1990s, however, taking photographs of oneself was not stigmatized at all, and had become a core activity among Japanese schoolgirls. New technology and the availability of inexpensive digital cameras, cell phone cameras, mini-Polaroids and disposable cameras were largely developed as a response to girls' interest in photography. Many girls began making photo diaries, scrapbook-like albums documenting everyday episodes and relationships. Although some critics attribute the popularity of this form to the success of Hiromix, the genre was already long established in girls' culture before it caught the attention of adult art circles through her work. As photography took root as an integral part of girls' creative output, it also became an ideal vehicle for the expression of defiant attitudes.

While many bad girl rebellions surface in a range of daily behaviors, we can also see resistance telescoped into the tiny spaces of a new form of photography called *purikura*. Coined from the name of one of the earliest coin-operated photo-editing machines, named Print Club, *purikura* are self-adhesive sticker photos. In 1994, a woman named Sasaki Miho formulated the concept based on her awareness of the popularity of sticker collecting and photography among schoolgirls. She offered the idea to her game company employers, but her male bosses didn't think it worth pursuing until 1995. The instant photo editing machine they finally created became an incredible commercial success as massive numbers of girls frequented these photo booths in trendy shopping areas, train stations, and game centers.

Non-scientific mini-polls in girls' magazines suggest that most girls visit a print club booth at least one a week, while some are more avid consumers. For one magazine article, twelve girls were interviewed about their *purikura* patterns, and asked about which machines they prefer, how many times per week they make *purikura* photos, and what they do with them afterward. The "print data" for frequency of use reported that five of them visited a booth two or three times a week, three of them had gone once or twice a week, one of them four or five times a week, while the remainder said they do it less than once a week.[10] Another poll asked one hundred girls "How often do you use a print club machine?" Almost seventy percent said they visit one weekly.[11] One girl was so taken with the activity that she dreamed of having her very own machine, claiming "If I have a *purikura* booth at home I'd be able to take them anytime I wanted."[12] Her commentary was accompanied by a comic making fun of *purikura* obsession: a teenager in her bedroom is running toward a print club booth planted in the middle of the floor, while her mother, looking in from the doorway, says "You're taking one alone *again*?"

Purikura are collected into albums, used for social exchange (sometimes called "thanks print club" or *okinipuri*), and for decorating everyday objects. They are so popular that *Popteen* magazine, a monthly targeting the under-eighteen crowd, uses them to adorn its horoscope pages. There are several companies that manufacture more than fifty different types of print club booths, and girls' media contain elaborate rankings, descriptions, and evaluations of the assets and qualities of each machine. Most consumers of print club are girls under eighteen. As they mature, however, some continue the practice, taking along children and spouses and including them in the activity. Girls have also been incorporating boyfriends and dates into print club culture, creating a sub-genre that is occasionally termed "love print club" (*rabupuri*). Having a *purikura* taken is usually a group activity involving two or more girls. (There are even instances of single print club stickers depicting fourteen girls.) The editing features are complex and sophisticated, allowing the photo-takers an opportunity to alter the image in numerous ways. The development of print club technology demonstrates a finely-tuned and close interconnection between girls' patterns of use and the developers' attempts to keep up with these.

Over time the construction of a *purikura* demanded a high level of cultural knowledge. Within girls' culture a photograph is considered bare and unfinished until it has been marked up with text and decorated with icons and drawings. In an attempt to attract more customers, print club engineers began adding text editing functions to allow graffiti and other annotations to be added before the photos are printed. Electronic pen features that affix outline style text or characters formed with glittering diamonds have proven to be quite popular. Girls' desire to experiment with script elements and types stimulates further development of photo technology, and these in turn affect the image-text that is produced, which leads to further technological fiddling and adjustment.

When girls are in the print club booth poised to anoint their photos with electronic pen markup, they have complete freedom to write and draw anything at all on them. This freedom inspires them to change a flat surface into an interactive object that often displays bad girl gender nonconformity.

Repulsive Photos

Speaking about *purikura* and their enormous popularity among Japanese schoolgirls, a foreign observer, interviewed by visual anthropologists Richard Chalfen and Mai Marui, related it to the "worship of cute" and said that:

> They present themselves as adorable, cuddly little things, desperate to preserve this one instant when they are genuinely cute. What better way than to accumulate a cute little notebook full of cute little images of themselves and their girl friends mugging into the camera and looking cute.[13]

Chalfen and Marui also see *purikura* as part of girls' consumption of cuteness, focusing on its smallness and its sweetly decorative aspects such as hearts and pretty frames. I would argue, however, that an examination of more recent *purikura* reveals forms of transgression that are distinctly uncute. Although many *purikura* continue to reflect aspects of the cute aesthetic, the form has also evolved in unanticipated directions. Even when the images are cute, the text may spoof it, as in cases of girls' labeling their own images with the graffiti "fake child" (*burikko*).[14] Provocative poses, slovenly or weird script and various defacements offer a counter to normative categories of gender, sexuality, and propriety that get challenged in the production of these photos. Bad girl self-photography and lewd graffiti are interesting precisely because they express so many subversive possibilities.

Purposely ugly photos are an obvious representation of bad girl resistance to codes of femininity. Girls sometimes call these "repulsive print club" or *yabapuri*, at other times they are categorized as "print club that gives you the creeps" (*kimopuri*), or as "dumb ass mode" (*aho mōdo*). *Yabapuri* show girls with eyes askew, fake blood dripping from noses, fingers or coins shoved into nostrils, food stuffed into gaping mouths, and distorted faces. Using the fingers to push the nose up to create a "pig face" is common. One photo of two girls shows them using their hands to manipulate the flesh of their faces, and one of them has pushed up her nose to approximate a pig's snout, while the other has pulled down her cheeks to create droopy eyes. They have written "most extreme weird face" on the photo. In another photo, two girls pictured with their tongues sticking out and pouting ferociously have written "disagreeable" in sparkling aqua characters beneath their unattractive faces. Often the words "pukey" or "nauseous" are added to similarly unlovely images, making it clear that they are calculatedly ugly. Explaining her love for one particular type of print club machine, one girl said that "You can seriously make a dumb ass face. You can really go over the top."[15]

The grotesque print club genre produces many cryptic and fascinating specimens. In a photo of two girls, they have added tiny dark green flecks to their lips, which are pursed toward the camera lens. The graffiti they wrote says "lips enchanted by dried seaweed," alluding to the seasoned seaweed customarily used as a topping on Japanese dishes such as fried noodles or octopus dumplings. In another photo, three girls are making grotesque faces, but they have also drawn huge antlers on top of their heads and have tagged it with "strange faces look like deer." As one *yabapuri* artist said "Just writing graffiti with a pen is already common, so I want to mark up my face as well."[16] In another photo two girls have drawn black moustaches, thick black brows, and black pipes on their faces, and have added the English word "Boss" to it. The text refers to a famous Boss Coffee advertisement featuring J-Pop star Hamasaki Ayumi, whose model-pretty face has been similarly defaced by the photographer.

Print club machines assist in *yabapuri* construction by offering a menu of icons and images that can be directly placed on the photo. Some machines have hundreds of "stamps" of animal parts, food items, and other unusual images. The Princess print club machine has 1,620 character sets and icons, while other machines allow two people to write graffiti at the same time. Another genre of the ugly photo is "horror mode" (*hōrapuri* or *hōra mōdo*), in which girls add icons from the world of popular horror iconography, such as streaming saliva, gushing blood, gory wounds, and various supernatural themes. In one photo two girls have added the characters for "ghost" and "curse" to their photo, in which they have overlaid bulging red-veined eyeballs and dripping blood onto their own faces.

The graffiti used on *yabapuri* is intertwined with the imagery to create obnoxious or offensive photos. A feature that points to girls' development of a deliberately bad script is to use the notation for the concept of "squared" in mathematics, which girls write as x^2 or ② to express intensity or reduplication. A photo that uses this pattern shows two girls enacting a strange scene. One is posed behind the other, reaching with her arms to catch an icon representing coiled feces, which is apparently coming forth from the buttocks of the other girl. At the top of the photo the phrase "it slipped out," is written, while descending down the side of the photo are scrawled the words "flopping shit" (*unko bitchi* ②). The photo is simultaneously funny and disturbing. Photo graffiti also manifests a clear influence from manga (comics) and the conventions comic artists have developed to indicate the texture of spoken language. A photo with two girls pretending to punch each other contains many effects shaped by manga's orthographic devices, with bursts that represent impact sounds or sweat marks that indicate extreme emotion or stress. The photo shows the girls engaged in a mock battle, and together with the words "big fight" they have added expletives such as "you bastard" and "damn!"[17] In another photo, two girls enact a make-believe molestation scene on a train or subway. They have drawn in hanging straps, which one girl is "holding" while the other is grabbing her from behind. The "molested" girl has sweat marks shooting off from her head, and "Pervert!" is written underneath.[18] A different ugly photo presents two girls acting like hungry animals. Their mouths are wide open with the teeth exposed in grimaces, and their hands are posed as if grabbing for something. The graffiti "Give us our grub!" makes the intention clear, since the word for food is one only used for animal fare.[19]

These intentionally grotesque photos are significant because they demonstrate that girls themselves have manipulated them through disfigurement and defacement,[20] seizing the right to do so before others can play with their images. As John Fiske noted when discussing resistance through popular culture "the everyday culture of the oppressed takes the signs of that which oppresses them and uses them for its own purposes."[21] Japanese girls are well aware of the commercialized and sexualized uses to which photographs of women have been put in mainstream and male-authored media, so they usurp this prerogative by altering their own photos at the point of manufacture. These ugly photos critique compulsory femininity and the oppressive emphasis placed on female beauty and cuteness through unsettling imagery. In addition to claiming the power to disfigure their own images, girls also express symbolic resistance through a form of gender vamping of sexualized themes.

Provocative Posing and Bad Girl Camp

In a cheeky print club photo of two girls, they have pulled up their school uniform skirts to reveal their underwear. The photo was sent to a magazine and published with this note from one of the creators: "If you look closely you can see hair sticking out."[22] Bad girl photographs such as this have become increasingly common, showing girls with clothing removed, posed in lascivious ways, and using various props and text in a suggestive manner. This genre of seemingly indecent self-photography became known as "erotic print club" or *eropuri*. In 2003 the weekly tabloid *Shūkan gendai* eagerly reported that girls were using print club machines to produce "raunchy" erotic photos. One commentator offered an analysis for why these new forms of self-display were popular:

> Girls use *eropuri* to create an image for themselves that encapsulates their dreams as being media personalities. Just as many girls imagine they're a rock idol every time they pick up the microphone to sing karaoke, they probably imagine they're some sort of glamorous photo model when they show off their bodies in a *purikura*.[23]

Although this interpretation seems reasonable, it is unlikely that many of the girls who make *eropuri* have this sort of conceptual model in mind. While their ostensible sexual themes and poses present them as objects of sexual appeal, there are aspects of the photos that confound or disrupt this interpretation. A typical *eropuri* as labeled by girls themselves often contains clues to its reading as spoof or caricature. The outrageousness of their bad girl posturing can therefore be seen as a challenge to sexualized femininity, not an affirmation of it.

The origin of *eropuri* is thought by some to reside in the development of a genre of photography known as the "self nude." Beginning in the mid-1990s, photographers such as Hiromix began taking and publishing shots of themselves unclothed, a trend denigrated as "girl photography." Nagashima Yurie's book of self-nude photographs in particular was seen as the inspiration for amateur girls' photography in a similar vein.[24] In addition to Nagashima and Hiromix, bad girl photographers such as Shirai Satomi, Miyashita Maki, and Nakano Aiko were also credited with pioneering the self-nude form. Miyashita's book of seventy Tokyo women photographed in their cluttered living quarters wearing their favorite bras and panties was grabbed up as erotica by some male consumers.[25] Other men, however, felt uncomfortable looking at obviously nonprofessional models photographed in such mundane settings by a woman, and wondered if perhaps Miyashita was a lesbian because she was interested in documenting women's lives, not appealing to men.[26] Although the self-nude genre is an interesting development in female photography, it is probably not the inspiration for *eropuri*.

It is most likely that *eropuri* has its genesis in the "costume play print club" (*kosupuri*). The concept of costume play in Japan has a broader meaning than it does as a borrowed term in other countries, where it generally means dressing up as anime, manga and video game characters. Japanese girls use the label for any type of masquerade. For example, a popular form of costume play is to dress up as a "race queen," the name for sexily-attired female promotional attendants at racetracks. During the late 1990s, game centers began to set aside areas for girls to deck out in costumes prior to stepping into the *purikura* booth to take their photos. According to one mini-poll, the most popular print club costumes include Stewardess, Race Queen, and Mini-skirted Policewoman.[27] Other preferred costumes are Santa, Chinese-style Girl, Nurse, and Bride. In one photo, five girls in Nurse uniforms have written "What about those huge boobs?" across the photo. The graffiti employs the vulgar term for breasts, yet juxtaposes this with feminine-marked language forms, producing a photo-text that is obvious satire.[28]

Some game centers that provided costume play print club excluded unaccompanied boys from entering, so groups of girls became bold in disrobing in front of the camera. Yet the disrobing is selective, and rarely results in full nudity. Usually it is bras and panties which are displayed for the camera, occasionally a bare behind or breast, or most common of all, manually-created cleavage. Girls say that showing your bra in an *eropuri* is no different from a photo of yourself wearing a swimsuit top. Girls rarely make *eropuri* alone, it is usually done in teams or groups. For example, four girls took a photo of themselves with their backs to the camera, their pants pulled down to display a few inches of backside cleft. The fact that the collective derrieres of four girls are lined up suggests that this is not intended to be an authentically sexy photo, but rather a type of coarse mugging. In another similar photo, two girls have photographed themselves from behind, their jeans pulled down a few inches. They have written "superior mooning" on the photo, as well as the characters for "white" and "black" on each girl, suggesting that they are members of competing teams, and that proper mooning is worthy of ranked evaluation. Commenting on the exposed bottoms seen in pictures in her old album, one girl added the cynical note "Thanks for the ultra nasty print club."[29]

Girls use *eropuri* to denaturalize sexualized presentation. In one print club the girls used the machine's "makeup" feature, which allowed addition of "cosmetics" to their faces, to create garish portraits. One of the photo-takers says they look just like "super cabaret women" (*chō kyaba-jō*), the term for sex workers found in bars offering low-end sexual services.[30] By describing their intentionally overwrought appearance this way the writer uncovers the constructed nature of gendered looks. Their cheerful vulgarity flouts pressure to aspire to the upwardly mobile and elegant style characteristic of good girls who wear subdued makeup and pricey clothing. In another photo that departs radically from the social expectation that girls must act modestly, two girls have taken a photo of themselves sitting on the floor facing the camera with their legs spread apart. The words "Looking is hateful!" and "nasty" are written on the photo, and the icons used on tourist maps to indicate the presence of hot springs are placed over their panties. The icons serve to draw attention to that which the text admonishes the reader not to view, while the scene depicts a taunting awareness of the male eroticization of the schoolgirl. Remarking on their choice of the hot spring icon, one of the girls says "For the stamp placed over the nether regions, the steam one is best!"[31]

Eropuri challenge the *Playboy* model of women as sexual objects instead of subjects through a type of self-proclaimed exhibitionism or boasting. Girls' evident awareness and manipulation of erotic conventions in their photos is contrary to expected norms for maidenly chastity and self-effacement. *Eropuri* mock the very visual codes that have been used to objectify them. Girls commonly write such things as "People good at sex," or "sexy shot" on photos of themselves. In one photo two laughing girls wearing bathing suits have pulled the tops down to expose their breasts, which are nevertheless re-covered with heart icons. The "look, but you can't see!" scene spoofs the concept of "natural" schoolgirl modesty and chastity. Girls often ham it up for the camera in ways that expose the artifice of sexualized gender. In another photo two girls simulate sex from behind, with one girl enthusiastically humping the other. The word "sex" is written on it in English, as is the Chinese character for "love." One of the photo's creators, with obvious tongue-in-check, says "We were overcome by sexual desire then. It was wonderful."[32] As Susan Sontag noted of camp, it "exists in the smirk of the beholder."[33] It seems that Japanese men rarely penetrate to this level of meaning in *eropuri*.

Critics in mainstream media view *eropuri* as a form of erotic adventurism, and as further evidence of the breakdown in morality among youth. Yet girls who produce *eropuri* are not inadequately schooled in gendered norms. On the contrary, they have learned them all too well and use *eropuri* as commentary on the way their culture sexualizes them. A photo ripe with theatricality includes two girls facing the camera with artificial cleavage they have created with their arms and by leaning forward. They are making stereotypical come-hither faces with pursed lips, and have written "sexy girls" in English underneath. In another caricature of the porn pose, two girls are likewise manually creating cleavage by scrunching their arms together under their chests. They have written "Big boobs? We don't have them so we made them!" on their photo. The accompanying commentary by one of the girls says "An illustration of us desperately creating cleavage."[34] In another similar photo of two girls, one is leaning forward with her chest toward the other, who is using her hand to pull the top out a bit and is peeking in. In addition to other requirements for female appearance, large breasts have recently become a focus of erotic objectification, a fact girls often mock through exaggerated displays such as this. Although Hollywood stars take their fake cleavage seriously, these girls certainly do not. When Japanese teens shove their breasts together and push them up for display in front of a camera, mocking the new hypermammary regard, they are doing it as a form of gender parody. The popular culture critic David Laing stated "An attempt to parody sexiness may simply miss its mark and be read by the omnivorous male gaze as the 'real thing,'" a comment that seems true in this situation as well.[35] Just as male critics miss the thrust of "girl photography" such as that created by Miyashita, they also fail to catch the vamping of gender intended in bad girl *eropuri*.

Bad Girl Photo Modalities

Deliberately ugly and provocative exhibition are not the only thing girls are inserting into their photos. They manipulate other resources to compose photo-textual ensembles that defy cultural norms, using objects and new script styles in their photos to assist in creating images which are unsettling or unseemly. The hedonistic pleasure they take in intertwining and manipulating text and image, usually seen as separate things, reflect playful and impertinent audacity. Their graffiti photo art form requires popular culture expertise and an ability to integrate textual wit, visual novelty, and bad girl panache.

The breakdown of rules for good girl gender display is seen in the actual script girls use on their photos, which may unnerve readers' expectations for written Japanese. Girls' writing contains many aberrant characters and subsidiary graphs. They reject the beautiful, orderly handwriting that is supposed to characterize female sensibility and refinement. One type of achievement for middle-class girls is thought to reside in the discipline and meticulousness required for mastering calligraphy. The misshapen and deviant writing found on girls' graffiti photos reveals opposition to expectations of standardized handwriting and its reification of aesthetic calligraphic styles. There is a widespread belief in Japan that good penmanship is an index of the proper background and character of the writer, so girls' use of unsanctioned script elements and malformed writing symbolically undermines gender socialization.

Girls have created many novel script innovations and styles that eventually surface in photo graffiti. They have a history of inventing their own rules for writing, and some older forms remain in their culture for decades. An example is the substitution of Japanese syllabic characters with Roman alphabet letters, a trend that first appeared in the 1980s and is still seen on graffiti photos. In one recent example of two girls making hideous faces, they have written "the cool crowd" or *kakkoii kei* on the photo, but have substituted the letter "E" for the Japanese syllabic characters for "ii."

In the late 1990s girls initiated a new practice in cell phone text messaging, referred to as "girl characters" (*gyaru moji*).[36] Girl characters include disarticulated Chinese characters and mathematical symbols, or Cyrillic letters used as substitutes for Japanese syllabic characters. Girls are awash in script overabundance afforded by computer symbol and font menus. It is a technological bounty that they exploit and embrace. There are two different substitution sets, one for each of the Japanese syllabaries, yet these are often mixed in actual practice. The Japanese syllable "ra" might be written with the Cyrillic letters Я (it represents "ya" in Russian) together with "a," or else the syllable "ko" with the mathematical symbol $\supseteq$. The girl character found most often in graffiti is the mathematical symbol $\cup$ for the syllable "shi."[37] Some girl characters represent words, such as using the symbol © to represent the address term *chan*, an endearing diminutive for "Miss." When girls play with their writing system in this way they are accomplishing two things. They are refusing to be the caretakers of beautiful calligraphy and are rejecting their role as custodians of "correct" language. Girl characters also extend the boundaries of what is considered the written Japanese language, challenging the notion of language as a unified shared system. By redefining the borders of linguistic possibility, girls are also demonstrating resistance to the uniformity and predictability of standardized print media.

Resistance to gender norms is also seen in the way girls use objects in the construction of their provocative gestalts of language-image. Food items, stuffed animals, accessories, and undergarments appear frequently as props in self-photos. Numerous photos show girls with bras placed on top of their heads or worn as outer accouterments. In one photo two girls wear white bras over their clothing, and pose facing the camera with their chests presented for inspection. The graffiti reads "Matching bras" and "C65" (referring to bra size, measured in centimeters in Japan). In another case, two girls photograph themselves wearing panties that cover most of their faces, with eyes peeking out of the leg holes. The attached note says "Undies song dance."[38]

Girl photography contests the model of girls as vapid consumers of all that is cute and dainty. Similar to the Hello Kitty vibrator, "cute" objects are used in an unchaste manner. What could be more adorable than a stuffed Winnie the Pooh bear? Pooh is ordinarily associated with innocence and sweetness, yet many times he is used abusively as a prop in girls' photo vignettes. In a cell-phone photo, Pooh appears positioned between a girl's legs, his bear snout directed at her crotch. The graffiti on the photo reads "cunnilingus Pooh."[39] In another photo, a close-up of Pooh is shown with a girl's hand buried in Pooh's crotch, the fingers appearing to rub the area there. The fondled Pooh has the words "in the middle of masturbating" written on it. Pooh is also used in more complex photo storytelling. Graffiti photos with an interactive narrative structure, in which comments or questions are matched with responses or assessments, are quite common. Instances of graffiti writing in which two or more authors share its context and construction are a form of text production rarely seen in other settings. In one photo, two girls enact a childbirth scene with Pooh seeming to come forth from between one of the girl's legs. The graffiti above the laboring girl says "when giving birth to Mr. Pooh," while the other girl in the photo is posed next to her with a speech bubble that says "give it your best Older Sister!" The bearbirth enactment also suggests a covert snigger at adult anxiety over non-procreating women and unwed teen mothers. A smaller stuffed bear appears in a photo of a girl who is forcing it into her mouth. The accompanying text, using so-called masculine language, says "I'm hungry so I'm gonna eat Mr. Bear!!"[40]

A print club photo may be the product of mass-produced technology, but it is most admired when it shows the hand of its individual creator. Girls value graffiti photos that show authenticity and creativity, and participate in critiquing them. An important aspect of graffiti photo culture is the evaluation of its form, language, and overall success in projecting a particular feeling or sentiment. Levels of analysis and meta-analysis are seen in the genre called "print club commentary" (*purikome*), which is published in magazines and on internet websites. Both amateur and expert girl commentators offer interpretations and assessments of the graffiti photo's semiotic strata. The critics judge the photo makers' proficiency in using the various technical features offered by print club machines, as well as their adeptness at melding artistic, linguistic, and stylistic skills into one photo. Individual photos that do not show adequate mastery are rated with no-nonsense appraisals such as "no good" (denoted with the acronym "NG"), or "it missed the mark." Tips for the woman photographer (she is sometimes called a "cameraman" with the English loan word) are often found in magazine articles that provide detailed information on posing, camera techniques, and editing suggestions. The makers of specific print club machines also appoint their own girl arbitrators who publish analyses on internet sites or in magazines. Now that *purikura* has matured as an art form, there are retrospective articles on girls' earliest efforts at making them. Osuka-chan, who works in a tanning salon, says of one of her "photos from the old days" that it has "a serious schoolgirl feeling."[41]

By exploiting multiple resources, girls are adept at stretching and amplifying the possibilities of the photo medium. Girls' script performativity intervenes in the orthographic rules of language, thereby destabilizing the concept of good writing and the boundaries and politics of language representation. The use of props, objects, and text results in scenes of unladylike gluttony, un-feminine language, and sexualized displays that poke fun at images of the prissy good girl.

Conclusion

An interesting use of *purikura* is to attach testimony as evidence of its truthfulness. In a magazine feature, girls talk about their dangerous and obnoxious male teachers.[42] Commentary such as "He always stinks like sake," "He tells us how to use condoms," and "He's into girls," are accompanied with the writer's photo, lending legitimization to what is written. Attaching a photo to one's personal opinion and publishing them is an unusual instance of private media breaking through into the public realm. Self-photos may have their origins as personal artifacts, yet they are increasingly found in mass media outlets.[43] Graffiti photos, *yabapuri*, *eropuri*, and other photo genres underscore a growing generational split and a change in attitudes. By inserting their private, ugly or "sexy" personae into the public sphere, girls are defying the cultural rule that women should display restraint, modesty, and self-effacement in public.

Bad girl photography compels our attention for numerous reasons. Grotesque and sexy photos are not just individual acts of defiance, they are part of a general transgressive foment that appears in many forms of girls' popular culture. Photography cordons off a safe space for girls to work though their ambivalence toward gender socialization and the media's sexualization of girls and women. Mainstream instructions and admonishments about proper adolescent femininity are countered by their image-text lampooning and their disdain for cultural rules about writing, modesty, and self-expression. The dialectic of complicity and opposition is forced into the photo's small square. We can't deny the commercial dimension of photo production, but the fact that it is tied to profit does not weaken or deflect its capacity for articulating girls' concerns. Bad girl photography is a form of visual argument, a strategy of cultural critique. Girls have seized creative control of the commodified female image, reclaiming the photo from its typical use as male arousal or pretty girl disciplinary. Whether we see girl photography as transgression for the new century or as the detritus of capitalist marketing, girls have taken a familiar form of gender commodification and used it for their own purposes.

Notes

I appreciate the support I received from Jan Bardsley, Masa Iino, and Scott Clark. All photo graffiti and magazine quotes were originally in Japanese and are translated into English unless otherwise noted. All translations are my own.

1. "New Cosmos of Photography" (Shashin shin-seiki) sponsored by Canon, 1995 Gallery Exhibition Report Online at <http://www.canon.com/scsa/newcosmos/gallery/1995/>.
2. David J. Sapir, "On Fixing Ethnographic Shadows," *American Ethnologist* 21, no. 4 (1994): 868.
3. Graffiti photos (*rakugaki purikura* or *rakugaki shashin*) are described in Laura Miller, "Graffiti Photos: Expressive Art in Japanese Girls' Culture," *Harvard Asia Quarterly* 7, no. 3 (2003): 31–42.

4. An overview of adult complaints about youth is found in Gordon Mathews and Bruce White's edited volume *Japan's Changing Generations; Are Young People Creating a New Society?* (London and New York: Routledge/Curzon, 2004).

5. The role of *purikura* in buttressing social solidarity is also noted in Richard Chalfen and Mai Marui "Print Club Photography in Japan: Framing Social Relationships," *Visual Sociology* 16, no. 1 (2001): 55–77. This friendship-marking function is seen in the earliest Japanese photographs: see the photo of two girls entitled "Girls of Good Friendship" (Ogawa Kazuma, 1860–1929), No. 345 in the Nagasaki University Database of Old Photographs of the Bakumatsu-Meiji Period, Online at <http://oldphoto.lb.nagasaki-u.ac.jp/unive/>.

6. For more on extreme makeup and fashion see Sharon Kinsella's chapter, this volume, and Laura Miller, "Media Typifications and Hip *Bijin*," *U.S.-Japan Women's Journal English Supplement* 19 (2000): 176–205. Impertinent linguistic forms used among girls are described in Laura Miller "Those Naughty Teenage Girls: Japanese Kogals, Slang, and Media Assessments," *Journal of Linguistic Anthropology* 14, no. 2 (2004): 225–247.

7. Anne Wilkes Tucker, Dana Friis-Hansen, Ryuchi Kaneko and Joe Takeba, *The History of Japanese Photography* (New Haven: Yale University Press, 2003).

8. Sakuma Rika, "Shashin to josei," in *Onna to otoko no jikū: Nihon no joseishi saiko*, ed. Kōno Nobuko et al. (Tokyo: Fujiwara Shoten, 1995–1998), 196. A painting showing three American men forcing a Japanese woman to pose for the camera is in Oliver Statler, *The Black Ship Scroll: An Account of the Perry Expedition at Shimada in 1854* (Tokyo: John WeatherHill, 1964), 56–57.

9. Frederic Scharf, Sebastian Dobson and Anne Nishimura Morse, *Art and Artifice: Japanese Photographs of the Meiji Era* (Boston: Museum of Fine Arts, 2004).

10. "Saikin purikura pōzu mihon," *Cawaii* November 2003: 45.

11. "Fotojenikku teku oshiechaimasu," *Cawaii* May 2003: 147.

12. "E-Girl no akogare mono," *Egg* 85 (November 2003): 87.

13. Chalfen and Marui, "Print Club Photography in Japan," 66.

14. This derogatory label is discussed in Laura Miller "You are Doing *Burikko*!: Censoring/Scrutinizing Artificers of Cute Femininity in Japanese," in *Japanese Language, Gender, and Ideology: Cultural Models and Real People*, ed. Janet Shibamoto Smith and Shigeko Okamoto (Oxford: Oxford University Press, 2004), 146–165.

15. "Purikura kishu chō tettei hikaku," *Popteen* June 2004: 197.

16. "Saikin purikura pōzu mihon," 145.

17. Ibid., 144.

18. "Purikura kishu chō tettei hikaku," 200.

19. "Saikin purikura pōzu mihon,"142.

20. It is impossible to determine the percentage of *purikura* that are deliberately ugly or naughty. Perhaps one fifth of those published in magazines such as *Egg, Popteen* and *Cawaii*, and on the internet are of this type.

21. John Fiske, "Cultural Studies and the Culture of Everyday Life," in *Cultural Studies*, eds. Lawrence Grossberg, Cary Nelson, and Paula Treichler (New York: Routledge, 1992), 137.

22. "E-Girl no puri chō Show," *Egg* 85, November 2003: 77.

23. Ryann Connell, "Bold Teen Babes Flash Full Bodyflesh for the Porn Print Pic," *Maininchi Daily News Interactive*, August 27, 2003. Online at <http://mdn.mainichi.co.jp/waiwai/0308/0827eropuri.html>.

24. Nagashima Yurie, *Nagashima Yurie* (Tokyo: Fuga Shobō, 1995).

25. Miyashita Maki, *Heya to shitagi* (Tokyo: Shōgakkan, 2000).

26. "Japan's Derided Women Photographers are Earning New Recognition," *AsiaWeek* August 13, 1999, Online at <http://www.asiaweek.com/asiaweek/99/0813/feat2.html>.
27. "Fotojenikku teku oshiechaimasu,"149.
28. The graffiti "*oppai dō nan no yo?*" includes female-marked and nasalized sentence final particles.
29. "Purikura guranpuri," *Ego System* 49, July 2004: 84.
30. "Purikura kishu chō tettei hikaku," 199.
31. "Yabapuri taishō happyō," *Cawaii,* January 2003: 148.
32. Ibid.
33. Susan Sontag, "Notes on Camp," in Susan Sontag, *Against Interpretation* (New York: Farrar, Straus, Giroux, 1966), 277.
34. "Saikin purikura pōzu mihon," 144.
35. David Laing, *One Chord Wonders: Power and Meaning in Punk Rock* (Philadelphia: Open University Press, 1985), 94.
36. Shibuya Hetamoji Fukyu Iinkai, *Gyaru moji heta moju kōshiki Book* (Tokyo: Jitsugyōno Nihonsha, 2004).
37. I doubt that secrecy is a motive for using *gyaru moji*. As evident in their use of spoken language, girls don't care if adults understand them or not.
38. "Ego *shame,*" *Ego System* 49, July 2004: 75.
39. "E-Jump photo mail," *Egg* 92, June 2004: 86.
40. The masculine-marked graffiti is "*hara hetta kara kuma-san kū zo!!*"
41. "E-Girls no puri chō Show," 74.
42. "Uchira no gakkō no abunai sensei," *Egg* 92, June 2004: 54.
43. "We'll show you our private and memorable print club," in "E-Girls no puri chō Show," 75.

Girl in a crowd. Photographed by Jeffrey Chiedo (2000). Used with kind permission.

9

Black Faces, Witches, and Racism against Girls

Sharon Kinsella

Between summer 1998 and summer 1999 *kogyaru* suntans began to get darker. The personality of the style veered from that of the slatternly coquettishness of dropout schoolgirls toward that of moody punk divas. Girls involved in this climactic phase of Shibuya, Center Gai street fashion used self-tanning crème and tanning salons to tan their skin as dark as they could, if possible to a chocolate brown color. Dark skin was highlighted with pearlescent colored eye shadow and lipstick, which, until the beginning of the decline of the look in late 2000, was used to paint thick white rings around the eye sockets and over the mouth. White-socketed girls redefined their eyes with dark eyeliner and false eyelashes cemented with lashings of mascara. The glamorous big hair of *kogyaru* style, streaked or dyed light red brown, made way for heavily highlighted whitish-blonde hair arranged in shaggy dos, and in some cases tonged and piled-up into bouffant arrangements. This powerful assemblage was overlaid with colors: metallic lame face glitter on the cheeks and around plucked arching brows; glittering face stickers in the shape of tear drops, stars and hearts; and equally well-encrusted fingernails and painted extensions. White-on-brown was accessorized with any of a range of generally theatrical props, from ubiquitous clusters of artificial tropical flowers strung on bracelets, necklaces, and hair slides; to colored contact lenses; temporary tattoos; cowboy hats; character merchandise and bulky ethnic jewelry.

In the press the terms "nega-film," "nega-make" (photo negative make-up), and "panda-make" were used to describe the facial expression. Girls became referred to unanimously as "black faces" (*ganguro*)[1] and girls sporting its most extreme affectations were called "witches" (*yamamba*). Racial innuendo joined, and to some extent displaced, the priapic innuendo ("loose socks" or loose sex?) paying court to *kogyaru* fashion. Three girls in particular, nicknamed Buriteri, Akoyoshi, and Fumikko, received brief media fame as the darkest witches on the streets. In addition to "black faces" and "black face girls" a range of hyperbolic temporary terminology, such as "mega-black" (*gonguro*) and "mega-girl" (*gongyaru*) was concocted to emphasize the tonal violence of the style. Interestingly the unflattering moniker *yamamba* is an antique term in current usage for the archetypical "mountain witches" or "hags" that appear in folklore and Nō theatre, and in ukiyoe woodblock illustrations of the plots of the latter[2] (see Copeland in this volume). In the male press the word *yamamba* in particular, embodied a barely disguised slur, which accurately reflected the common editorial sentiment of abusive animosity toward this self-involved and ostensibly frightening stage of *gyaru* fashion. Except in jokes and parody, "witch" (*yamamba*) was not the word chosen to describe girls' style within *kogyaru* magazines like *egg*, which became specifically dedicated to the radical and tanned look from 1999.

Weekly news magazine headlines and television report anchor men reacted in tones of exaggerated horror, but even a cursory backward glance through the decades of girls' comics, literature, theater, and fashion magazines, demonstrates that girls' culture and fashion in Japan has been riddled with wayward racial affiliations and pseudo-ethnic expressions throughout its propagation from the early twentieth century. What is more, a rummage through both near contemporary and historical writings and social policy on girls and young women, illustrates that rather than being a novel concern, maintaining a national stock of racially pure, sexually chaste, and ethnically Japanese young women, and then protecting them from the damaging temptations of foreign travel, foreign female behavior and fashion, and racial miscegenation, has been a longstanding concern. A complex antiphony has evolved between ideological, literary, and aesthetic proscriptions of virginal, obedient, gentle, and maternal *ideal* Japanese girls, emanating almost entirely from the educated male camp, and what might be called the "anti-Japanese" tendency of girls' culture. Throughout the many genres and forms of girls' culture, both subtle and theatrical expressions have been generated of dynamic girl characters with invented hybrid ethnicities.[3]

The much commented upon aura of sexual inexperience and purity which underpinned prewar *shōjo* and postwar cute cultures respectively, was in reality a posture with both sexual and racial coordinates. Virginal prewar girls' culture and asexual and individualistic postwar cute culture have, with some minor exceptions, been implicitly bourgeois, European and white, in orientation. Prewar *gāru* (and *modan gāru*) culture and contemporary *gyaru* culture emerging from the 1980s, through which is bridged a certain continuity, have been characterized, by contrast, as assertive, brazenly sexual, and oriented toward exotic, urban, and tourist locations, and white and black American music and style, from jazz to hip hop.[4] If the eventual value of the exaggerated postures of either untouchable guileless virginity, or overbearingly frank and precocious sexuality, has remained unresolved, and if girls' culture in Japan has been cleaved accordingly into two main streams rooted in the different habitus of middle and working class life and employment, the element which has remained constant across and throughout the different modes of girls' culture, and which serves to articulate it, finally, as a single tide, has been its consistent turning away from the trappings of traditional Japanese femininity, ethnicity, and idealized female Japanese looks.[5] Young women displaying their enthusiasm for either the closeted spheres of girls' culture and communications, or more cosmopolitan styles of female behavior have, in turn, been singled out and stigmatized as racial and cultural traitors to Japan. The continual surveillance of girls' mores and fashion by an eagle-eyed "male press" (*oyaji zasshi*), that has taken upon itself the task of charting and disciplining signs of feminine evolution and contrariness, has simultaneously provided a rapt national audience and receptive stage for entertaining cultural digressions undertaken by the more brave-hearted of young women.

Little Girls (*kogyaru*), Witches (*yamamba*), and Black Faces (*ganguro*) in the Media

Articles distributed in 1999 and 2000 protested that black faces and witches were an affront to the tastes of male readers. "Big Survey of Aesthetic Taste: Teenage Witch Girls Should be Worried!" warned *Spa!* magazine. The *Weekly Jewel* demanded "We Want to See the Real Faces of Our Black Face Daughters!"[6] The same slough of weekly magazines targeted at male readers, that had connected radical girls' fashion to casual prostitution earlier in the mid-90s, now complained that black faces and witches were trying to sell themselves but were repelling male customers. "Cabaret Clubs Have Become Lairs for Those Ugly Witches" grumbled the *Weekly Post*, while *Focus* magazine protested, "Are We Going To Have Even More of These Witch and Black Face Porno Videos!?"[7] These articles framed their judgment and damnation of this particular girls' street style in terms of an unequivocal sexual rejection.

Though rooted in the wily rump of the self-consciously male press (e.g. *Shūkan bunshun, Shūkan post, Shūkan gendai, President*), caustic derision of black faces and witches became a prototypical position enthusiastically taken up by other sections of the public. *Ganguro* was received less as style than as cultural "travesty."[8] Leading female artist Tabaimo has portrayed a schoolgirl in uniform squatting to defecate on the national flag (*Japanese Zebra Crossing*, 2000). During fieldwork observation carried out in winter 1999, Toshio Miyake, noted that "More and more of these girls flaunt themselves, regroup on the streets, and adopt provocative attitudes, by which they expose themselves to verbal abuse from passersby, physical violence, the prurient winks of older men, and getting headhunted by scouts working for the sex industry."[9] In an article published in the respectable organ *Bungei shunjū* and thought suitable for translation and abridgement for the *Japan Echo*, one female writer ridiculed the risible aesthetic faux pas committed by black faces and witches. "In all honesty" she confided, "I have seen very few girls sporting the style that brings me even close to thinking, 'Without that makeup, she must be a beauty, what a waste.' "[10] Pursuing this attack, Nakano Midori suggested that stupidity was the key to the style: "Nothing about it is pretty, elegant, or stylish; the main effect, I would say, is to frighten. These girls almost seem to be wearing placards that say, 'I'm stupid.' Meeting someone who so overtly insists on her own idiocy tends to scare people. It overpowers them." The allegation that witches and black faces were ugly *and* stupid, circulated widely and formed a base stereotype, underlying more intricate considerations of their hygiene and racial origins: "From *Kogyaru* to Witches, Platform Boots, Black Face, Idiot-ization: *Kogyaru* On the Darker and Dirtier Program."[11] On television shows much play was made of "moron black faces" (*ōbaka no ganguro*) and taciturn specimens were filmed replying to probing questions from anchor men with the single monosyllable "..*eeeh*" ("I dunno..").

Several photographic projects on *ganguro* carried out around the turn of the century, seemed to share a similar instinct to present black faces and witches as pitiful, and déclassé. In Ōnuma Shōji's portrait of the black faces of the summer of 1999, the viewer is invited to see the disheveled and lopsided appearances of the girls' faces beneath their bedazzling first appearances.[12] Ōnuma focuses on unflattering details: the way in which tan foundation crème is sliding off hot oily skin, or the way in which skin rashes can be seen protruding through layers of lame glitter. These imperfect surfaces seemed to imply that, rather like Impressionist portrayals of French prostitutes, *ganguro* is a style soaked in the aura of cheap and failed glamour. Another series of enlarged facial portraits of black face girls without their make up, taken by young female photographer Sawada Tomoko, was exhibited at the *Futuring Power* Cannon photography competition held at the Tokyo Photography Museum in September 2002. Sawada's large, fine-grained pictures show six pudgy adolescent faces with slightly unfocused and confused gazes. Each face has pimples, badly plucked eyebrows, and blotchy, discolored skin. Among other things, the photographs seem to suggest to the viewer that *ganguro* girls using tanning salons to change their appearances are dimwitted young creatures engaged in an egotistical folly.

Yamamba and *Ganguro* as Primitives and Animals

Interpreting the brown skin cultivated first by *kogyaru*, and subsequently pursued to extremes by *ganguro* and *yamamba*, provided the occasion for a particularly perverse squall of journalistic pontification on the zoological, racial, and ethnic origins of girls. Rather than reading black face as an intelligent style, as a clearly deliberate instance of sartorial communication (à la Hebdige),[13] it was merrily misinterpreted as a form of animal coloring or tribal decoration. Girls who could not afford tanning salons were said to be using oil-based magic markers for eyeliner and coloring in their faces with dark brown marker pens.[14] An irreverent vein of reportage in the male press adopted a mock scientific tone and colonial language to claim that radical girls were a kind of species prone to natural selection. The magazine *Modern* (Gendai) for example, presented: "Professor Kashima Explores the Heisei [1989–] Jungle in Search of 'Uncharted Regions of Everyday Life' 3: 'Platform Boot Witches' No Longer in the Lead in Shibuya."[15] The notion that the energy and desire associated with *kogyaru* and black faces, was in some way primitive and animalistic, circulated around men's magazines and around girl's magazines themselves. Freelance writers, researchers, and contract editors were an important vector shunting ideas between readerships representing different sections of society with quite different attitudes. Several of the major and minor *kogyaru* magazines for example,—*egg*, *Popteen*, *Street Jam*, *Happie*, were produced by editors, mostly men, previously engaged with making pornography for men, in several cases in adjoining offices of the same company. One freelance female writer specializing in producing articles about *kogyaru* for the male-centered press and television, as well as working with *kogyaru* magazines for teenage girls, confidently imagined that "they are like primitive people who don't use words or language or books, people who just exist by means of images, their appearances, and their body adornment. If they want something they just take it, they are material animals, they are not interested in culture or society, they are only interested in money."[16] Another article in the liberal weekly *AERA*, described the sexual exploits of the "Terrifying Drunken Tiger Girls."[17] In other articles a connection was insinuated between black face girls and witches, and Africans or Southern people: "Is it the Influence of Global Warming, Evolution, Or a Passing Trend? Probing the 'Latinization' of Japanese Youth! Witch girls in monster make-up, lax about time and appointments, kissing and arguing in public, relaxed about sex."[18]

Smug references to the skin color, lifestyle, and possible ethnicity of black faces and witches bled into one another in a manner that illustrated the continued co-mingling, at least in popular journalism, of anthropological ideas about culture and biological conceptions of race. For tanning their skin and adopting new attitudes, hair color, and clothes, girls were indiscriminately accused both of African mimicry and in fact of being, or becoming, tribal, primitive, black, or a new ethnic breed. As Jennifer Robertson has remarked upon, these types of essentially Lamarckian ideas about the possibility of acculturation into a racial way of being were quite typical of prewar racial consciousness internationally. In the Japanese case "race" (*jinrui*), and "ethnic group" or "people" (*minzoku*), were, and in the context discussed here, continue to be, viewed as largely interchangeable concepts.[19] Furthermore, commentary about the race, tribe, and skin color of girls, was sometimes entwined with a derogatory and pseudo-Darwinian commentary about dark-skinned girls, which implied that they were a kind of species or animal.[20] Classified as dark-skinned primitives and animals, girls daring to wear black face and witch outfits sometimes became subject to a racist assault on their humanity.

The previously mentioned photographic portrait of black face girls by Ōnuma Shōji is titled *Tribe (Minzoku)*. In a short afterword by Tad Garfinkel, the girls are variably described as primitives and animals: "Like all the animals walking on the continent of Africa they have their own style. Just like Giraffes and Ostriches. Shibuya is a Safari! They shout out loud and clear 'We are a tribe!' Well done! That's right! You are the Japanese gypsies."[21] A review of this book posted on the website of the *Gendai nikan* newspaper suggests that it is a photographic testimony of "a sudden change in *kogyaru* DNA that lead to the birth of a new subspecies of the Japanese race (*minzoku*)."[22] Less explicit intimations that *kogyaru*, black faces, or witches, could be approached as a kind of jungle-dwelling tribe of anthropological interest were present in the widespread technique of presenting "uninitiated" readers with labeled anatomical line drawings of girl specimens, and elaborate vocabularies of girls' slang presented as a foreign language. Comic artist Koshiba Tetsuya included an explanatory anatomical diagram of his lead character and a list of "*kogyaru* terminology" on the inside back cover of collected volumes of his popular men's comic series about a *kogyaru*, *Tennen Shōjo Man* (Wild Girl). Another extensive vocabulary of *kogyaru* language (much of which appeared on closer examination to be comprised of preexisting slang terms in wider public circulation), was published in the sedate older man's magazine *dacapo*.[23]

The editorial of the men's trend watching magazine, *Dime*, invented the term *gyanimal* to describe "girl-animals" in an article titled "*Gyaru* + Animal = Gyanimal Breeding." The article proposes that girls wearing animal prints, gold lame, metallic fabrics, and other brightly colored items, were trying to attract and snare men.[24] An insert column by a specialist of girls' cultures suggests that in his opinion, "this fashion is very similar to an animal rutting season," in that, "lipstick in wine red color is in vogue, and that is precisely the same color as the vagina of a female monkey on heat." Positioned among this animal behaviorist commentary is a full-length photograph of a model decked out as a *gyanimal*. On the next page the model is stripped of all trace of self-tanning cream and animal-print micro skirt, and this pale and plainly dressed incarnation, most closely resembling a polite office lady, is presented as an *anti-gyanimal* and *Dime* editorial's own "ideal girl." On the next page, writer, Mori Nobuyuki (author of the *Tokyo High School Girl Uniform Fieldbook*) makes the only slightly less risqué suggestion, that *kogyaru* fashion comprises a collective "warning color, which, like the bright markings of tree frogs, say to potential predators 'I have poison. Eating me is dangerous!' "[25]

An innovative article about *ganguro* and *yamamba* fashion published in the *Weekly Playboy* applied a mixture of Darwinian theory, Native Ethnology (*minzokugaku*), European colonial fantasy about Africa and jungle primitives, and contemporary politically correct ideas about the social inclusion of ethnic minorities.[26] Pithily titled "Witch Girls Must be Classified as National Cultural Property Before it is Too Late," and subtitled "Is There a Danger of Shibuya Street Girls Becoming Extinct?," the writer intimates that the girls are a kind of ethnic minority, which may, like an endangered species of animal, "become extinct." The article is accompanied by a pyramidal diagram titled "The Shibuya Hierarchy," which illustrates in ascending order the evolutionary stages, from *gyaru* at the bottom, through *gangyaru* and *gongyaru*, to *yamamba* at the apex, who are presented as a kind of dark skinned female über race, reigning over earlier evolutionary forms. In this diagram, gender difference literally shades into racial difference. *Weekly Playboy* goes on to argue that by pursuing black identity, black faces and witches have arrived not so much at a semblance of contemporary black culture, as at the primary stage of human evolution, which is rooted in Africa, and is based on the principle not of money but of "black magic." However, making a case for the enlightened tolerance of this primitive girls' ethnic group in modern Japan, the article ingeniously cites "an African think tank" which has calculated that "in view of the falling birth rate, in order for Japan to maintain its current level of economic development in the twenty first century, it will have to admit up to six million foreign workers a year." This article captures the imaginative association of primitive African tribe, the native folk of Japan, and contemporary girls' culture, which are elided into one continual formation. The writer concludes that: "As Japan entered modernity it underwent homogenization. Holding dear the illusion that homogeneity = good, Japan lost the ability to activate the people ... the *yamamba* may be a warning to Japan. Will the girls' culture be protected or will it be eliminated? The future of Japan rests on this question."[27] *Playboy*'s ham statement that the future of Japan is bound up with coming to terms with the ethnic status of Japanese *ganguro* girls is considerably less preposterous than it might at first appear. The precedent for this intriguing dissolution of female sociology into female ethnicity was established both in semi-academic analyses of girls' culture produced from the mid-1980s, and in popular portrayal of girls in art, animation, and culture. Moreover the deeper logic underlying this cultural imagination rests on twentieth century theories and feelings about a quite distinct Japanese race or people, whose survival hinges upon the successful sequestration of pure-blooded and dedicated young Japanese mothers.

Girls' Studies

What might be tentatively considered a new sub-discipline of Girls' Studies emerged in a number of books published from the mid-1980s, which sought to investigate the concept and lifestyle of *shōjo* and *gyaru*. Girls' Studies was concerned with explaining contemporary girls' cultures, such as cuteness, or the so-called *gyaru* subcultures, *bodikon* (body-conscious) and *oyaji gyaru* (mangirls), of female college students and office ladies. With the exception of the work of Honda Masuko, editor of *Girl Theory* (*Shōjoron*, 1988),[28] Girls' Studies was pioneered by male scholars[29] and tended to analyze contemporary girls in the context of national history and culture. Widely read works in this little oeuvre include: Ōtsuka Eiji's *Native Ethnology of Girls*, 1989; Yamane Kazuma's *Morphology of Girls' Handwriting*, 1989; Honda Masuko's *The Alien Culture of Children*, 1992; Yamane Kazuma's *Structure of the Girl*, 1993; Masubuchi Sōichi's *Cuteness Syndrome*, 1994 and Kawamura Kunimitsu's *The Body of the Maiden*, 1994.[30] Girls' Studies demonstrate a thematic convergence between contemporary Cultural Studies and Native Ethnology or folk studies, a field incorporating aspects of religious studies, psychoanalysis, and cultural anthropology.[31] Hovering between academic analysis and popular non-fiction writing, the majority of these pop-ethnologies function less as descriptive academic studies and more as the *ur* texts of cultural professionals, journalists, and *otaku* ("obsessive fan") critics.

In 1989 a young journalist trained in cultural anthropology and connected to, what at the time was still a largely underground network of reclusive young men producing Lolita complex (*rorikon*) subculture, published a book about the mysterious nature of Japanese girls. The main argument of Ōtsuka Eiji's *Native Ethnology of Girls: End of The Century Myths About The 'Descendents of the Miko'* is that there is continuous anthropological lineage from the ancient *miko* shrine maidens to the cultural rituals of contemporary teenage girls. Ōtsuka describes aspects of girls' culture of the 1980s as a tribal or ethnic system of culture and connects contemporary girls to Yanagita Kunio's "common people." Ōtsuka postulates that through the transformation of a rural peasant society into an urban consumer society: "Modernity has changed the Japanese folk (*jōmin*) into girls (*shōjo*)."[32] The logic of this thesis is that active, unmarried, urban young women, a group that has historically represented a toxin to holistic national ideas predicated upon a pure and traditional Japanese femininity, can be effectively collapsed back into Japan. A unifying rusticity has been located in the ritual behavior of urban girls, so that, rather than the countryside, girls themselves have become the vehicle of a mysterious living nativism in the midst of the city.[33] Honda Masuko also reminds readers of the ancient practice of female shamanism, in a poetic treatise about the magical and aesthetic qualities of girl children.[34] Honda describes girl-hood as "the quivering" (*yureugokumono*)—an aesthetic trace of a "different world" that is not absolutely real. Girls, Honda proposes, are complicit in their own outsider status and the segregation of girls' aesthetics and pastimes from the rest of modern culture. Incarcerated in schools and dormitories, girls are other-worldly beings that are implicitly foreigners: "Theories of the everyday order can not even formulate the words required to discuss this gypsy-like sensibility."[35]

Between 1981 and 1984 a cram school student waiting to re-sit his university entrance exams carried out fieldwork on schoolgirls in uniform at one hundred high schools in and around Tokyo. *The Tokyo High School Girl Uniform Fieldbook* (*Tokyo Joshikō Seifuku Zūkan*), which became a classic text of Lolita complex sub-culture, was described by well-established *otaku* critic Nakamori Akino, as an example of "cultural anthropology," neatly demonstrating, most of all, the inter-weaving of academic social sciences in knowingly low-brow, male entertainment. Mori, however, chooses to contradict Nakamori, and states that in truth his inspi-ration came from his boyhood fascination with illustrated picture books about birds, fishes, or insects. The humor upon which the book's entertainment value rests is its deadpan categorization of schoolgirls as a species of naturally occurring national fauna. Akasegawa Genpei, a ubiquitous figure of the postwar avant garde, jokes in an enclosed review that he "had realized that high school girls in Tokyo were breeding. But I had not realized that they constitute a separate species."[36] The reactionary attitude of the book toward girls gained critical attention from unexpected quarters, when The Japan Uniform Manufacturers lodged com-plaints against the *Fieldbook*, which they claimed "treats schoolgirls as objects."[37] Allusions to schoolgirls as animals in mass formation crop up in early academic studies of girls' and in contemporary film (e.g. Sono Shion's *Suicide Club* [*Jisatsu circle*], 2002) and in the visual arts (e.g. Aida Makoto and Matsukage Hiroyuki's *Gunjōzu* [*Ultramarine-scape*], 1997).[38] In his essay, *Girl as Subject*, Kohama Itsurō suggests that the cliquey habits of girls are essentially those of "pack animals" (*guntai dōbutsu*), who "exhibit their eroticism not as individuals, but as a solid collectivity."[39]

Meanwhile in the *Structure of the Girl*, freelance scholar, Yamane Kazuma noted that changes in girls' behavior during the 1980s, led them to drink, smoke, and begin walking about on the streets at nighttime. Bold girls began to meet foreign-ers in nightclubs and to gather in Roppongi on "streets that brimmed with state-less power."[40] Yamane compares *gyaru* of the 1980s to the less Teutonic races of the Southern hemisphere: "The active mode of girls today is similar to that of Latin people in the South. The figure of a *gyaru* in a disco, clad only in a mini-skirt, a tight-fitting outfit, or even literally half-naked, sweating as she dances furiously away, suggests scenes from the Rio carnival. Southern people are extremely cheer-ful, happy-go-lucky and hedonistic. Sexually liberated too, they act almost as if they had never experienced suffering. Southern people thoroughly enjoy their lives and Japanese *gyaru* today are beginning to proximate the culture of the South."[41] Yamane goes on to suggest that as a country in the Northern hemisphere, Japanese society is correspondingly governed by the erstwhile European and Protestant principles of "industriousness" and "self-denial." In the midst of this industrious society, girls' culture was revealed as an alien element, as a "Southern race" within.

Female artist, Mariko Mori, picks up the theme of the Japanese girl as post-modern national shaman in her photographic portrayals of a mystic native place, centered about the presence of sacred girl characters. *Nirvana*, a 3-D animation presented at the Venice Biennale in 1997, featured ex-model Mori posing as Amaterasu, the Goddess of Japanese creation, seated within a computer animation of a lushly-colored primal Japanese landscape. In another animation, *Shaman Girls' Prayer* (*Miko no Inori*, 1996), Mori, wigged in white and donning white contact lenses, proposes herself as a futuristic Japanese female creature with shamanic powers, which allow her to interact telepathically, with the advanced technology of Osaka International Airport. Girls transformed into mythological Shinto spirits, *miko* shamans, and rustic maidens in kimonos, have became exceedingly common in boys' and mens' comics, animation and computer games, such as Takahashi Rumiko's *Inuyasha* (serialized in *Shōnen Sunday*, serialized from 1996) or Samura Hiroaki's *Blade of the Immortal* (*Mugen no Jūnin*, serialized in *Afternoon*, serialized from 1994). In Miyazaki Hayao's animated films, little girls are the heroic defenders of ancient Japanese tribes and their lands. Aspects of the rural arcadia, common folk and mysterious animistic characters of Yanagita's earlier writings seem to reemerge in Miyazaki's fantastic stories. *Princess Mononoke* (1997), for example, is a wolf-child who wears a red mask with markings and a cape of white fur attached, during her battles with armies invading the countryside. In this oversized mask, a white tunic and a dark blue skirt, *Mononoke* most closely resembles a Japanese schoolgirl dressed as a tribal primitive. Equivalently success-ful in communicating to another international milieu, artist Aida Makoto has con-tinually returned to the image of a schoolgirl as a key symbol of his nation. In a painting entitled *Azemichi* (Paddy Path, 1991) Aida presents the back view of a schoolgirl in sailor uniform walking between rice paddies. A central parting divid-ing the girls' hair into bunches, forms a vertical line at the center of the painting, which is continued into the line of the footpath she is walking. The girl traverses and is incorporated into an archetypical site of traditional or native Japan.

Dark Skin, Race, and National Purity

The dual and interchangeable categorization of girls, as either the saviors of Japanese folk culture and national ethnicity, or as an unpleasant alien racial sub-presence within the nation, illustrates the previously indicated, continued mental proximity of cultural ideas of ethnicity and lifestyle, with scientific ideas about biological races. Scientific racism came to dominate the social and natural sciences of Europe and America during the same decades in which the Meiji government sought to import modern Western learning to aid Japanese enlightenment and militarization.[42] Meiji intellectuals, such as the preeminent Fukuzawa Yukichi, subscribed to the theory that humanity was arranged in a natural hierarchy, in which yellow people occupied a middle position, and black and dark skinned people, occupied the bottom position, next to apes.[43] The cycle of association between yellow and brown skin, human primitives, and apes, strengthened through the prewar and wartime period, in both Japanese cartoons of its Asian neighbors and colonial subjects as dark-skinned, and sometimes fat-lipped and unintelligent too,[44] and in the "simian image"[45] of Japan itself, which became ubiquitous to wartime coverage of the Japanese in Britain and America. While the relative inclination toward ranking and characterizing race according to skin color has fluctuated according to other political affiliations, and over the duration of Japanese colonial expansion, occupation, and recovery,[46] John Russell suggests that the simple notion of black people as an ape-like and subhuman species, which gained an early root in the modern Japanese imagination, was still in circulation in late postwar popular culture.[47] Nakasone Yasuhiro's infamous comments in 1993 about a "mongrelized race" problem weakening the moral cohesion and work ethic of the United States,[48] also illustrate that blackness and signs of so-called racial mixing continue to be associated in certain powerful circles, with antisocial and subhuman behavior. Those most closely associated with black people and culture in postwar Japan have been women and girls working as prostitutes, and wayward young women choosing to identify with black American culture.[49]

Not withstanding particularly invariant and fetishistic characterizations of people of African descent, conviction in the fuller idea of a racial hierarchy determined by skin color was understandably ambivalent, and often muted, within Japan in the twentieth century. Rather than skin color, theories of the Japanese race (*yamato minzoku*) developed in the Meiji period and expanded through the prewar, centered upon blood and sexual reproduction. Maintaining the purity of the "bloodline" (*kettō*) of the nation, primarily through the continuous interbreeding of racially and culturally pure Japanese, has positioned the sexual and reproductive activity of young Japanese women at the center of national racial defence.[50] Jennifer Robertson reports that the "central focus of the Japanese eugenics movement concentrated on the physiques and overall health of girls and women," who were perceived anew as "the biological reproducers of the nation." The ongoing program for the protection of the reproductive maternal body, and the stigmatization and racial rejection of young women appearing to flirt with foreign cultures or engage in sexual relationships with non-Japanese men, are conjoined facets of the tendency to manage female sexual reproduction. Defensive "ethnic national endogamy"[51] required Japanese girls to dedicate themselves to their future Japanese husbands alone, making virginal schoolgirls the natural and enduring counterparts and mythological partners of heroic young kamikaze pilots setting off on their missions during the Pacific war.[52]

A eugenic program that regarded Japanese girls as the bodily vessels of national ethnicity, regarded hidden, unlicensed, or casual prostitution (of which "compensated dating" [*enjo kōsai*] is the contemporary correlate), as the main vector through which unsuitable racial mixing might take place. Through the system of licensed prostitution under police surveillance, prewar governments sought to segregate chaste and preblooded Japanese girls and mothers, from those working in the brothel trade. In her work on colonization and female sexuality in Imperial Japan, Sabine Frühstück has argued that the use of incarcerated Asian females, many of school age, as well as overseas Japanese prostitutes, as comfort women in Japanese military brothels, "was an extreme form of the colonization of sex and was closely intertwined with debates about and practices of the control of prostitution in civilian society at the time."[53] Prostitutes bearing features considered the signs of racial purity were assigned to have sex with a higher rank of Imperial soldier, and vice versa. Government fears in August 1945, that Japanese womanhood would be raped and impregnated indiscriminately by the imminently arriving Occupation army, or that they might become the "concubines of Blacks,"[54] informed the rapid assemblage of special brothels dedicated to American servicemen stationed in Japan.[55] Impoverished and often homeless young women were invited to serve the nation by volunteering themselves to what was conceptualized as a "blockade" of prostitutes' bodies, providing sex to foreigners and thereby heading off the threat of generalized racial mixing.

Despite government attempts to enforce national objectives, Japanese women expatriated from military brothels in China and Korea, domestic prostitutes, and young women stranded without a means of survival, flowed on to the streets of destitute, defeated Japan and began conducting business for themselves.[56] Girls who slept with white and black American soldiers were nicknamed *panpan*, and became an emblematic figure of early postwar society, treated with both fascination and contempt for their "bold and subversive" opportunism.[57] Liaisons between young Japanese women and American soldiers preoccupied the prurient and painfully emasculated male imagination of the early Occupation period, forming the blueprint for a traumatic conflation of libertine women with national military defeat, and the threatening presence of a foreign (sexual) power.[58] In particular a "classical association"[59] developed in postwar imagination between black men, blackness in general, and prostitutes. John Dower has commented that "race hates did not go away," after the Pacific war, "rather they went elsewhere."[60] One place where traces of a racial system of thought have resurfaced in the ongoing social war inside postwar Japan is in the discourse about youth, from "tribes" to the "new breed," and in the distinctly racialized characterization of "yellow cabs,"[61] "black faces," and other fashionable young women.

Anti-Ethnic Girls

Though stimulated to a fever in journalism bating girls decked out as black faces and witches at the end of the 1990s, the entertaining innuendo that *ganguro* and *yamamba* girls were in fact a primitive tribe or species of animal, was not entirely novel, nor separable from the ethnic terms of analysis of semi-academic studies of girl's culture. Serious and tongue-in-cheek commentary on girls as animals and dark-skinned aliens predated the fashions adopted by some teenage girls in the 1990s by several decades.

In the flamboyant polycultural tastes of *kogyaru*, and the dark-skinned, white-lipped, blue-eyed mischief of *ganguro* and *yamamba*, the trajectory of female cultural imagination and experience, which has crystallized around an ongoing ambivalence toward traditional culture, reached an explosive stand off. It was the more remorseless of the critics who appeared to appreciate the experiential origins of *ganguro* and *yamamba* style most precisely. Says one writer who preferred not to beat about the bush: "The effect is such that it makes me want to ask, 'Are you a prostitute from some foreign country, or what?' " [62] Rocked in the cradle of a society literarily and literally dominated by male cultural and intellectual production, girls' street fashion, managed albeit, by young, culturally informed, and hip magazine editors sympathetic to what has been coyly referred to as "girls' feminism," [63] secreted a silent, stylistic response, which caught up, echoed, contradicted, confused, and incited the barrage of male journalism and broadcasting, peremptorily accusing girls of sexual and racial delinquency. Radical girls' style is demonstrably rooted in the same ideological framework as that of its critics, and responds closely to the ethnocentric preoccupations of discourse about young women. To use the words of an early British deviancy theorist: "the latent function of subculture is this—to express and resolve, albeit magically, the contradictions which remain hidden or unresolved in the parent culture." [64] Peculiarly racialized sartorial gestures worked to a baroque acme by *ganguro* and *yamamba*, and in unnamed future forms, constitute an intimate and knowing reply, to the fearful and reactionary fantasies about the dangerous and exotic behavior of girls, allowed to saturate national communications during the 1990s.

Notes

1. "Black face" is translated as two words, so as not to conflate it directly with the American term "blackface." See Nina Cornyetz, "Fetishized Blackness: Hip Hop and Racial Desire in Contemporary Japan," *Social Text* 41 (Winter 1994): endnote 2.
2. The *yamamba* is a mountain witch of prodigious strength who lives as a bitter recluse in the mountains. Her superhuman power was often made available to assist men. See Mariko Tamanoi, *Under the Shadow of Nationalism: Politics and Poetics of Rural Japanese Women* (Honolulu: University of Hawai'i Press, 1998), 122. The *yamamba* has been adopted as a proto-feminist figure by some women, such as the novelist Ohba writing in the 1970s. See Minako Ohba's story, "The Smile of the Mountain Witch," in *Stories by Contemporary Japanese Women Writers*, ed. Noriko Mizuta Lippit (London and New York: M. E. Sharpe, 1982), 182–196.
3. Miriam Silverberg argues that the prewar modern girl "who was both Japanese and Western—or possibly neither—played with the principal of cultural or national difference. Seen in this way, she highlighted the controversy over adoption of non-Japanese customs into everyday life and called into question the essentialism . . . that subordinated the Japanese woman to the Japanese man." In "The Modern Girl as Militant," *Recreating Japanese Women 1600–1945*, ed. Gail Lee Bernstein (Berkeley: University of California Press, 1991), 245, and 263. Laura Miller documents the signs of cultural hybridity in *kogyaru* fashion in "Media Typifications and Hip Bijin," *U.S.-Japan Women's Journal English Supplement* 19 (2000): 176–205; and "Youth Fashion and Changing Beautification Practices," in *Japan's Changing Generations: Are Young People Creating a New Society?*, eds. Gordon Mathews and Bruce White (London and New York: Routledge/Curzon Press), 83–97.

4. As E. Taylor Atkins records, consciousness of the black social roots of jazz music were muted in prewar Japan. Jazz cafes and dance halls were nevertheless linked to a suspected collapse in female sexual morality. *Blue Nippon: Authenticating Jazz in Japan* (Durham, NC: Duke University Press, 2001), 121–123, 110–111.

5. Read more about the complex interaction of "cross-dressing and cross-ethnicking" (132) in girls' theatre in the colonial period in Jennifer Robertson, *Takarazuka: Sexual Politics and Popular Culture in Modern Japan* (Berkeley: University of California Press, 1998), 89–138.

6. "10dai yamamba gyaru osoru beki bi-ishiki dai chōsa!!" in *Spa!* (September 1, 1999), 136; and "Ganguro musume no sugao ga mitai!" in *Shūkan hōseki* (April 14, 2000), 54.

7. "Kyabukura wa yamamba mitai busu no ni natta," in *Shūkan post* (October 8, 1999), 63; and "Tadaima AV ni mo zōshokuchū ganguro, yamamba tte ii!?" in *Focus* (March 8, 2000), 24.

8. Toshio Miyake, "Black is Beautiful: Il Boum Delle Ganguro-Gyaru," in *La bambola e il robottone: Culture pop nel Giappone contemporaneo*, ed. Alessandro Gomarasca, (Torino: Einaudi, 2001), 111–144.

9. Miyake, "Black is beautiful."

10. Midori Nakano, "Yamamba," *Japan Echo* 27, no.1 (February 2000): 62–63.

11. "Kogyaru kara yamanba e: Atsuzoko, ganguro, bakkaka. Kogyaru wa shidai ni kuroku, kitanaku," in *Spa!* (July 1, 2003): 26.

12. Ōnuma Shōji, *Minzoku* (Tokyo: Kawade Shobō Shinsha, 2001).

13. Dick Hebdige, *Subculture, The Meaning of Style* (London: Methuen, 1979).

14. "Ima koso yamamba gyaru mukei bunka sai ni shite," in *Shūkan Playboy* (May 2, 1999), 198.

15. "Heisei jungle tanken—Kashima kyōju, 'nichijōseikatsu no hikyō' o motomete kyō mo iku 3," in *Gendai* (February 2002): 326.

16. Uchida Kaoru, interview in the Hitotsubashi publishing district (November 8, 1997).

17. "Toragyaru osorubeki enjo kōsai: Joshikōsei saisentan rupo," in *AERA* 9:16 (April 15, 1996), 62.

18. "Ondanka no eikyōka? soshite ichiji no boom ka? shinka ka? Nippon wakamono no Latin-ka genzō o saguru!" *Spa!* (February 9, 2000), 47.

19. Jennifer Robertson points out that "Like their international counterparts, Japanese eugenicists tended to collapse biology and culture, and, consequently, held either explicitly or implicitly Lamarckian views on race formation and racial temperament." See "Blood Talks: Eugenic Modernity and the Creation of New Japanese," in *History and Anthropology* 13, no. 3 (2002): 196.

20. The frequent elision of black and simian imagery until as late as the 1980s, is discussed by John Russell in "The Black Other in Contemporary Japanese Mass Culture," in *Contemporary Japan and Popular Culture*, ed. John Treat (Richmond: Curzon/ University of Hawai'i Press, 1996), 24.

21. Ōnuma, *Minzoku*, 2001 (no page numbers).

22. *Gendai nikan* online at <http://www.bookreview.ne.jp/list.asp>.

23. See "Shōjotachi no shingo, ango, ryūkōgo," in *Dacapo* (October 15, 1997): 88. Researchers Maruta Kōji and Fujii Yoshiki likewise found little evidence of a "schoolgirl language" and conclude that it was a fiction of the mass media. See Maruta Kōji, "Giji-ibento to shite no enjo kōsai," *Osaka jogakuin tankidaigaku kiyō* 30 (2000): 210. Laura Miller examines the controversy over girls' language in "Those Naughty Teenage Girls: Japanese Kogals, Slang, and Media Assessments," *Journal of Linguistic Anthropology* 14, no. 2 (December 2004): 225–247.

24. "Gyaru+animaru= gyanimaru zōshoku," in *Dime* (January 1998): 10–11.

25. *Dime* (1998: 11). Sociologist, Miyadai Shinji, criticizes the tendency to caricature schoolgirls as a species in his analysis inflected by rational choice theory. (*Seifuku shōjo-tachi no sentaku* [Tokyo: Kōdansha, 1994], 283.)

26. *Shūkan Playboy* (May 2, 1999): 198.

27. Ibid., 201.

28. A fine-grain, monochrome photograph of a naked pubescent girl posed against a black background is featured on the cover of *Girl Theory* edited by Honda Masuko (Tokyo: Seikyūsha, 1988).

29. For a closer examination of one strain of male cultural investment in girls see: "Fantasies of a Female Revolution in Male Cultural Imagination in Japan," by the author, in *Zap-pa (Groupuscules) in Japanese Contemporary Social Movements*, ed. Sabu Kohso and Nagahara Yutaka (New York: Autonomedia, 2005).

30. Ōtsuka Eiji, *Shōjo minzokugaku* (Tokyo: Kōbunsha, 1989); Yamane Kazuma, *Hentai shōjo mōji* (Tokyo: Kōdansha, 1989); Honda Masuko, *Ibunka to shite no kodomo* (Chikuma Gakugei Bunko, 1992); Yamane Kazuma *Gyaru no kōzō* (Tokyo: Kōdansha,1993); Masubuchi Sōichi, *Kawaii shōkōgun* (Tokyo: NHK Shuppan, 1994), and Kawamura Kunimitsu, *Otome no shintai* (Tokyo: Kinōkuniya Shoten, 1994).

31. Marilyn Ivy, *Discourses of the Vanishing* (University of Chicago Press, 1995), 66.

32. Ōtsuka, *Shōjo minzokugaku*, 246.

33. Marilyn Ivy looks at related connections forged between urban young women and native Japan in the Discover Japan advertising campaign. See *Discourses of the Vanishing*, 29–65.

34. Honda's preliminary essay on this subject in *Shōjoron* (1988) is expanded in *Ibunka to shite no kodomo* (1992).

35. Ibid., 180–181.

36. Akasegawa Genpei in Mori Nobuyuki, *Tokyo joshikō seifuku zukan* (Tokyo: Kuritsusha, 1985), 208.

37. Mori Nobuyuki interview, Ryōgoku, Tokyo (March 20, 2003). Mori's approach bears a family resemblance to the ethnographic diagrams generated by urban folk studies or modernology *(kōgengaku)*, pioneered by Kon Wajiro during the 1920s. Interestingly, Kon himself apparently noted a similarity between his own method of intensive visual observation of his subjects, and that used by "botanists and zoologists." From Harry D. Harootunian, *History's Disquiet: Modernity, Cultural Practice, and the Question of Everyday Life* (New York: Columbia University Press, 2000), 186. Mori's zoological taxonomy of schoolgirls exploits the dehumanizing potential of this older disciplinary ambiguity.

38. Descriptions of girls as a "numerous and undifferentiated pack, devoid not merely of humanness and individuality," were, in common with accounts of their primate-like behavior, somewhat reminiscent of wartime racial stereotypes propagated in Allied media, of the Japanese per se. See John Dower, *War Without Mercy: Race and Power in the Pacific War* (New York: Pantheon Books, 1986), 93.

39. Kohama Itsurō, "Shutai to shite no shōjo," in *Shōjoron*, ed. Honda Masuko (Tokyo: Seikyūsha, 1988), 98–97.

40. Yamane, *Gyaru*, 60.

41. Ibid., 61.

42. Dower, *War Without Mercy*, 204.

43. Russell, *The Black Other*, 24.

44. Dower, *War Without Mercy*, 210.

45. Ibid., 86–87.

46. Ibid., 218–219.

47. Russell, *The Black Other*, 19.
48. Dower, *War Without Mercy*, 315.
49. The actual and the fantastical relationship of young women with black American soldiers in the Occupation period became a self-conscious theme of feminine photography (e.g. Yoshida Ruiko's *Hot Harlem Days*, 1967) and fiction (e.g. Ariyoshi Sawako's *Hishoku* [Colorless], 1967), by the 1960s. And, as Nina Cornyetz has documented in "Fetishized Blackness" (1994), hip hop attracted clusters of girl fans through the 1980s.
50. Robertson, "Blood Talks," 198–199.
51. Ibid., 192.
52. Dower, *War Without Mercy*, 232. Ironic references to virginal schoolgirls and young soldiers continue to crop up throughout avant garde genres. For a contemporary parody of the protagonist archetypes, see artist Aida Makoto's comic story *Mutant Hanako* (Tokyo: ABC Shuppan, 1999). Read Linda Angst encountering this mythology in Okinawa in "The Sacrifice of a Schoolgirl: The 1995 Rape Case, Discourses of Power, and Women's Lives in Okinawa," *Critical Asian Studies* 33: 2 (2001): 243–264.
53. Sabine Frühstück, *Colonizing Sex: Sexology and Social Control in Modern Japan*. (Berkeley: University of California Press, 2003), 41.
54. John Lie, "The State as Pimp: Prostitution and the Patriarchal State in Japan in the 1940s," *Sociological Quarterly* 38: 2 (1997): 256–257.
55. John W. Dower, *Embracing Defeat: Japan in the Wake of World War II* (New York: Norton /The New Press, 1999), 126–130.
56. Sheldon M. Garon, *Molding Japanese Minds: The State in Everyday Life* (Princeton University Press, 1997), 197.
57. Dower, *Embracing Defeat*, 132.
58. Joanne Izbicki discusses both the perceived impotence of Japanese men and the simultaneously overt sexualization of Japanese women in "The Shape of Freedom: The Female Body in Post-Surrender Japanese Cinema," in *U.S.-Japan Women's Journal English Supplement* 12 (1996): 109–153.
59. Nina Cornyetz, "Power and Gender in the Narratives of Eimi Yamada," in *The Woman's Hand*, ed. Paul Gordon Schalow and Janet Walker (Stanford University Press, 1996), 444.
60. Dower, *War Without Mercy*, 311.
61. Aki Hirota criticizes the rumor popular in the Japanese media that Japanese girls abroad are so easy to pick up, that they have been nicknamed "yellow-cabs" by English-speaking foreigners; see "Image-makers and Victims: The Croissant Syndrome and Yellow Cabs." *U.S.-Japan Women's Journal English Supplement* 19 (2000): 83–121.
62. Nakano, "Yamamba," 2000.
63. Enthusiastic, that is, about the emergence of active girls, but less interested in organized opposition to institutional sexism. See Ōtsuka Eiji, *Etō Jun to shōjo feminism-teki sengō subculture bungakuron* (Tokyo: Chikuma Shobō, 1998).
64. Phil Cohen, "Subcultural Conflict and Working-class Community," in *Working Papers in Cultural Studies* 2 (Birmingham: CCCS, 1972), 23.

A Filipina–Japanese couple, christening their son, surrounded by their friends and an "American" priest, 2001. Photo by Nobue Suzuki.

10

Filipina Modern: "Bad" Filipino Women in Japan

Nobue Suzuki

Yes, I asked my aunt [in Japan] to bring me there. My mother [also in Japan] told me that it was fun to work at a bar. I kept saying, "I wanna go! wanna go! wanna go!" Well, the thing is, Filipinas going to Japan all seem to come back beautiful. . . . Their skin has grown light. I also wanted to look like that, right? I wondered how they became so beautiful? Well, [I figured that] I should go after all and I'll become pretty too.

Julie, in her early twenties in the early 1990s, thought going to Japan offered a sure way to acquire the kind of modern appearance that many Filipinas her age desired.[1] Faced with financial difficulty, she also decided to go to Japan to earn money to enable her younger sisters to continue their schooling. She convinced herself by saying, "Find some means to raise money. School, school, school. Money, money, money! I'm the eldest and I must be strong!" In the end, Julie went to Japan, got a job as a bar hostess, worked hard, and saved. But she did not give all the money to her family. She gave about 20 percent to her parents and kept 80 percent for herself. After marrying Masaki, a Japanese public employee, in 1995, she began investing in property in the Philippines. As a sign of her middle-class life, human value, and security, she now owns a fancy 4.8 million-peso ($12,000 @ $1 = PhP25) condominium near the foreign embassies, overlooking Manila Bay. For Julie, working and marrying in Japan enabled the realization of her multiple goals of being simultaneously filial and modern.

Asian women's migrations to the West and intimate associations with Western men have often been considered ways they liberate themselves from the oppressive "traditions" of their Asian homelands. On the other hand, tales about Filipino women (Filipinas) such as Julie in liaisons with Japanese men are not so liberating: popular allegories commonly depict them as desiring not so much affective relationships but material gain. To achieve this goal, they are thought to take up jobs in the "sex industry" in Japan and marry Japanese men through meetings at nightclubs or matrimonial agencies. Filipinas like Julie who work at bars are thus suspected of using sex to achieve their unsavory ends. By crossing the prescribed borders of gender, class, and nation, they are seen as sexual and moral transgressors who are contrasted to the ideologically valorized wife-mother of the home within the national boundary of the Philippines and Japan. This assumed transgression from women's proper place and financial power has made these Filipinas "bad girls" in both countries.[2] Simultaneously, activists and academics portray Filipina immigrants as victims of gendered, classed, and nationalized North–South power disparities where "patriarchal" men of "rich" Japan dominate women of the "poor" Philippines. Filipinas in Japan have been seen as yet to achieve the modern life of the First World. Some Western feminists would conversely view these Filipinas "bad" because they ironically reinforce the desirability and power of the modern state and its men, instead of challenging them.[3]

The majority of such work on border-crossing women, however, disregards the multifaceted realities of their lives, which are continuously created under changing social and historical conditions. In particular, they fail to embrace the women's subjective views of migration abroad and work or marriage with foreigners. To assume the essential features of Filipinas' migration to Japan renders these women impersonal, faceless objects of (non) study, an objectification that prevents an understanding and appreciation of the women as multidimensional historical subjects of their own lives.[4] While various forms of surveillance do subject women, Filipinas themselves generate new meanings and reorganize their affective and material relationships in their lives.

Drawing on the data collected through interviews and participatory observations of the lives of Filipina wives in the Tokyo area in the 1990s, this chapter problematizes these academic and popular discussions by inquiring into some of the ways in which Filipinas attempt to attain modern identities in the course of their migrations and marriages to Japanese men. In the burgeoning literature, modernity is understood not as capitalist rationality and development or individual interests as defined and led by the West since today such capitalist modes of social organization are found globally. Modernity in this chapter instead refers to people's imaginations and attitudes about contemporary global realities and achieving self-realization.[5] Thus, while Westerners and Japanese make people of the "poor, traditional South" a reference point to lionize their own ethnocentric imaginaries of progressive difference, the Filipinas described here too make prolific attempts to attain their own versions of modern dispositions and lifestyles by reworking a range of discursive and lived differences in their everyday lives situated in a global terrain.

Filipinas' modernity in their daily lives with Japanese husbands is better captured in the spaces uniquely created between Japan and the Philippines in their differential historical relations to the West, notably the United States. Filipinos have been strongly influenced by the colonization by Spain (1565–1898) and the United States (1901–1946) and continuing American capitalist and cultural presences in the Philippines. Despite the criticism that Filipinos have been suffering from a "colonial mentality," recent scholarship argues that Filipinos' gestures and language mimicking America—through for example, performing American pop culture and speaking English—are ways to bring its symbolic power into their daily lives.[6] As mimicries, this "America" is not real, yet serves as a rhetorical device for Filipinos to imagine and believe that they behave *like* Americans. As shown in greater detail, "Americanness" allows Filipina wives to position themselves and their self-claimed "mono-cultural," monolingual Japanese husbands onto an East–West axis, where the women are powerful "Westerners." In so doing, the Filipinas transform colonial power into a key medium in negotiating gender and national hierarchies in their marriages in Japan. The women's reference points are thus not two dimensional but are multi-sighted, flexibly hovering across conventional divisions and symbols of power.

Hence, the notion of the Filipinas' "badness" I elaborate in this chapter is emergent and *not* inherent in the women, their motivations for migration and marriage to Japanese, or their culture and "Third World-ness." My discussion centers on two interrelated foci of Filipinas' migration to and marriages in Japan: first is the ways in which these women have been constructed as "bad" due to assumptions about their lack of choice and morality *as* women of the "South." Second, by exerting their agency Filipinas make themselves "bad" by both transgressing stereotypes and articulating their own dispositions and imaginations in their relations to such conceptions and to their husbands. Moreover, the women's multidirectional visions in making their own lives disturb the pernicious polar divides between the "deprived and backward" South and the "developed" North. The chapter thus details the women's ongoing ways of being modern, or of relating to differences with Japanese men, and of (re-)capturing subjective meanings in their own lives, lived across national, cultural, and gendered borders. I begin with the ways in which Filipinas have been narrated as transgressors.

Filipinas Narrated

Since the 1990s, Filipinos have composed the fourth largest group of foreigners in Japan—after Koreans, Chinese, and Brazilians—numbering approximately 170,000 in 2002, of which 84 percent were women.[7] During this time, Japan has become the second largest destination after the United States in Filipinas' marriage migration. As a result, among all registered Filipinas in Japan, an estimated 90,000 (57 percent), are current or former spouses of Japanese nationals, and the majority of Filipina–Japanese couples live in urban regions.[8]

The presence of Filipinas in contemporary Japan caught popular attention beginning in the 1980s and academic reports began to appear from the 1990s. Although migratory flows between Japan and the Philippines during the past century are complex, Filipinas in contemporary Japan have been primarily discussed in light of the influx of "*Japayuki*" (Japan-bound) entertainers working in night entertainment businesses in urban centers and of "*hanayome*" (brides) married to rural men. Most academic observations are methodologically problematic, showing a tendency to rely uncritically on politically and morally charged (Christian) activists' accounts of Filipinas who sought help and were willing to talk; on secondary materials, many recycled journalists' publications;[9] and on the number of apprehended workers and other speculative estimates.[10] This propensity has engendered teleological narratives of Filipinas in Japan that have inevitably revolved almost exclusively around the dark and difficult sides of their experiences.

Of course, representations of Filipinas are not monolithic or unchanging.[11] The two prevailing images have nonetheless haunted Filipinas' lives in Japan. Although entertainers work as hostesses whose main tasks usually are to converse with customers, serve drinks and food, light cigarettes, sing, and dance, many observers have insisted that the category "entertainer" is a euphemism and that such women (are forced to) work as "sex workers," often involving gangsters.[12] Other writers point out the illegality of hostess work done by entertainer visa holders since hostessing is not stipulated in their visas.[13] Though there have indeed been cases in which prostitution has been forced upon women workers in Japan, not all Filipina entertainers engage in prostitution or are coerced by gangsters. Many former "talents"—as they call themselves—with varied degrees of musical and conversational skills told me that they were not so much concerned about the illegality of work as the strict immigration control that curtails their desires to travel to and work in Japan.[14] The other prevalent image of Filipinas is that of "hanayome." These wives have been considered victims because they are "imported" with money paid by their husbands, who had been suffering from being "unable to marry," and because the women were "compelled to marry anyone" in order to fulfill their daughter role for their families in the Philippines.[15] Even 20 years after these wives' first arrival in the mid-1980s, observers continue to call them "hanayome" (or "new brides"), a term usually referring to women who are about to be or have just gotten married. The women have thus been discursively circumscribed within this gendered category, with its expected role as one of the lowest positions in traditional households in Japan.

The seemingly opposed images of "prostitutes" and "hanayome" nevertheless converge at the intersection of the Filipinas' gender, class, and nationality. The narratives underscore that many of these Filipinas are eldest daughters who are culturally considered "second parents." While this role is indeed important for both Philippine institutions and eldest daughters' personal identity, as Julie attests above, in such narratives the women are often rendered automaton role performers devoid of creative thought or any ability to transform their life conditions.[16] Beneath these allegories is the unexamined belief about women from the Third World being universally paupers, although numerous studies have shown that international labor and marriage migrants are not the poorest of the poor.[17] Sellek, for example, writes, "Prostitution is a common form of employment for [Southeast Asian] women attempting to *eke out* a living in urban areas. . . . Young, *innocent* women from impoverished countries in Asia enter Japan to work . . . in the sex industry . . . [and] to improve their [own and their families'] social standing through marriage to Japanese men."[18]

Underlying these pernicious representations is a tacit contrast of Third World women's "troubled" lives and the academic/activists' own privileged lives. The frequent equation of the *Japayuki* with the *Karayuki* (Japanese prostitutes during the prewar period) and "comfort women" (forced military sex slaves during World War II), without considering the historical specificities of their experiences, magnifies this difference between "us" and "them." It locates the Filipinas in a uniformly traditional and impoverished past,[19] far from the modern world and affective relationships western scholars implicitly claim to enjoy. While concentrating only on "sex workers" and "*hanayome*" that invoke the audience's curiosity and ideologically shaped moral concerns, even well-meaning scholars and advocates have neglected to investigate the messy but thus far invisible aspects of the Filipinas' quotidian experiences, where things may develop unexpectedly.[20] Consequently, some Filipinas have come to feel that skewed representations are products of "the media blitz on the plight of Filipino women [living and] working in Japan as bar hostesses [and wives] that was wrecking havoc on [the] Philippine image in Japan."[21] To take seriously this kind of critique raised by a Filipina wives' group in Japan, I now turn to some of the reasons, beyond economic need and filial duty, Filipinas decided to leave for Japan.

From Duties to Desires

> There were lots of Japanese coming to Manila. I got used to seeing them. Cassette-radios, cameras, color TVs, and those things. . . . I thought there were money-bearing trees in Japan . . . When you go to the black market . . . the dollar and yen are really strong. America is far away. So, naturally Japan is the place [to go for work].[22]

As this comment made by Dinah, a 19-year-old Filipina in the mid-1980s shows, despite widespread (but not universal) poverty in Philippine society, ordinary Filipinos' imaginations about affluent lifestyles and modern amenities are not impoverished. This statement may be hastily read as evidence of "Third-World" people's cravings for riches and comfort abroad. However, Filipino historical contexts complicate such reasoning. Since the 1960s and increasingly after the overthrow of the Marcos authoritarian regime in 1986, capitalisms in Japan and the Philippines have caused ordinary Filipinos to become producers and/or consumers of an unprecedented amount of Japanese goods and technological products. Filipinos have also received massive amounts of Japanese popular culture, along with business and tourist visitors. Through these, Filipinos have come to imagine the wealthy, modern life available in Japan.[23] Successful return migrants from Japan are further assurance of this possibility. Rather than being alienated from the products and services that cater to the Japanese, many ordinary Filipinos have begun pursuing their desires for those modern goods and lifestyles. Thus, Julie, as quoted at the beginning of this chapter, decided to work as a talent at nightclubs in Japan. In her case, Julie was fully aware of her family responsibilities as the eldest daughter, but she also wanted to attain the material, the fun, and the beauty afforded/able by such work. Through this, Julie fantasized her new identity in a convivial global center beyond the compass of Philippine society. Other Filipino migrants to Japan similarly conjure up such possibilities. The following is the case of Marissa.

Marissa was born in the early 1970s, as the third of six children. In the late 1980s, when she was still in college, many problems—such as accidents of family members and the unplanned birth of the youngest sibling—occurred one after another. Marissa's parents were working, but their earnings were not sufficient for the family. Her older and married siblings could support only their own families. Under these economic and family circumstances, Marissa wanted to help her parents and so decided to undergo dance training in preparation for work in Japan. Marissa, like many other Filipinas I know, did not tell her family about her intention to go to Japan until a few days prior to her departure. She knew the horrifying rumors about entertainers forced into prostitution and about gangster operations. Her grandmother, who lost her husband to the Japanese Army during World War II, opposed her wish to go to Japan. However, by insisting that she would work hard to enable her younger siblings to continue schooling, Marissa persuaded her worried elders. As she was traveling toward the airport, this 19-year-old woman "imagined the shuttle bus to be like my airplane" and then she saw the real plane. Marissa felt bubbles of excitement popping up in her heart, "Wow, cool!" As soon as she boarded the plane, she was "thrilled to go to famous places like Tokyo Disneyland!" After arriving in Japan, she also "enjoyed meeting the Japanese and speaking with those who could converse in English."

Although Marissa talked about her family's financial difficulties and her wish to fulfill her filial duties as the responsible daughter remaining at home, her story also revealed a teenage woman's excitement about air travel abroad. Despite her awareness of potential dangers, Marissa wanted to experience the pleasures of a modern and middle-class lifestyle. Such desires actually began even before she left home and her wishes were intensified when the manager of her entertainer promotion agency first took her dance company to the nearby American military base to perform there. When Marissa moved out from her strict parental supervision, that was typical for unmarried girls and women in the Catholic Philippines, she was thrilled about this excursion, saying, "All the ways of thinking were 'American' style! I felt like I was already in America [giggle]!" However, Marissa was not totally enticed by everything American. She said, "I did not like Americans because, this was a bit bizarre story, but 'theirs [penises] are big,' everyone said [vociferous giggle]. So, I was scared! . . . At that time, I liked Americans who were only a bit small. But the manager didn't allow us to go out without his chaperoning."

Philippine scholars have argued that Filipinos are extremely interested in thinking of themselves in reference to America and variants such as the "West," "foreign," and "modern," as the source of power, wealth, beauty, and leisure.[24] As seen in Marissa's case, "America" serves as a trope of self-empowerment and an access point for situating oneself in global realities. So, Marissa was afraid of direct (sexual) contact with "big" Americans, but wanted someone physically small and so psychologically manageable. She saw her trip to Japan as offering such an "America" in the numerous symbols of the modern—the air conditioned shuttle bus, Tokyo Disneyland, and English-speaking people (however imperfect Japanese abilities might actually have been) on the way to and within Japan. Given the increasing visibility of Japan in the Philippines, Dinah too saw Japan to be an easy substitute when America appeared to be too distant. By going to Japan, Julie thought she could transform herself into a beauty with fair skin, a symbol of the upper class. In these explorations of "America" these women remain in charge of their wishes, rather than being driven by necessity, to achieve modern, "American" identities. Similarly, Filipinos' mobilization of the American trope also becomes extremely important in spousal negotiations. With this and other devices, Filipinas further express their desires for "modern" affective relationships, which in turn destabilize their otherwise negatively narrated identities.

Spousal Negotiations: *Kamakake* and Other Ploys

The prevailing narratives about Filipinas married to Japanese tell us that they were "acquired" in the "bride trade" and that the women alone are subjected to patriarchal power in their relationships in Japan.[25] As shown below, some women do suffer from hardships, but many Japanese husbands of Filipinas told me how their wives tricked them with *kamakake*, a sort of "white lie," to check the men's love and commitment to their relationships before and in marriage. Securing a man's commitment in marriage is vital for Filipinas because in the Philippines the law does not recognize divorce, and marital status remains integral to women's identity. Broken relationships with Filipino men are one serious reason why many Filipinas decided to leave the Philippines and marry Japanese men.[26] Stories about Filipina talents being single mothers have been well circulated among the male customers of Philippine pubs in Japan. During courtship, as is consistent with their gender construction as family providers, many Japanese men worry whether or not their fiancées have children in the Philippines for whom the men may have to become responsible. This anxiety is exacerbated when the men first visit their fiancées' families in the Philippines, where dozens of children may be running around. Filipinos' general tendency to be physically affectionate even with other people's children further confuses men. While picking up and rubbing their cheeks against a child's, sometimes it is hard to figure out at a first glance who the child's mother is.

In this context Rita, a former talent in her late twenties, ran her fiancé Terami through a *kamakake* trial. When he visited Rita's natal family, a grinning Rita pointed to a child and told him, "Hey, look! That's my kid!" Terami's immediate response was, "Ah, wouldn't I know it?" (*yappari nē*). Aware of his knowledge about this "possibility" and about Filipinas' familial roles, Rita turned these well-circulated tales about Filipina "entertainers" to her advantage. Invoking the popular discourse and her fiancé's gender construction, Rita tested Terami's response. Though real single mothers were not always successful in keeping their relationships after introducing children to their suitors, Rita was able to confirm that Terami was ready to accept her and "her child." By receiving this kind of response, Rita also brought Terami closer to her social world. Rita's ploy was, however, mild. Other women have checked their suitors through more radical *kamakake*.

Julie met her husband, Masaki, at the bar where she was hostessing. While at the bar, her workmates married to Japanese were telling her that after the nuptials Japanese men often became "cold." Julie knew also about the marital breakups of Filipina–Japanese couples, including that of her aunt. Julie felt compelled to examine Masaki's commitment to their relationship through a stronger method than Rita's *kamakake.* Prior to their wedding, Masaki proposed to meet Julie's family in the Philippines. Seeing this as an opportunity to figure out his character as well as to let him know about her personality, she and her mother decided to "do something fun for our guest coming all the way from Japan!" So, they took Masaki to a strip show with the intention of testing his manly ability to stand up to temptation. Upon the women's request, a naked dancer toyed with Masaki, a 230-pound strong, black-belt judo wrestler, with her breasts and crotch. Masaki froze, his eyes locked, and glued his hands to his thighs while Julie and her mother were intently watching his reactions. Following this ordeal, Masaki was deemed not to be promiscuous, a serious cause of marital dissolutions in the Philippines. But the game was not over. In the days following, Julie continued her trial of Masaki by intentionally blowing air and sitting in such a way that her panties showed. After all this, Julie judged that he would make a good husband committed to their relationship no matter how she behaved. Once Filipina–Japanese couples have gone through these prenuptial trials, they enter negotiations as spouses. "America" again comes into play in their struggles over different qualities of Filipinas and Japanese in a deterritorialized, transnational arena.

Inserting "America" into the Marital Mix

As discussed above, Filipinas' embarkation for work abroad can be led by the desire to have a close relation with an imagined "America." Filipinos, for example, sing American songs in English, which in turn allows them to bring the power of "America" into their everyday lives. On the other hand, Japanese more commonly relate to America as being a contrasted, external other. Emphases on contrast may be used to underscore their nation's uniqueness,[27] but are also sources of ambivalence and sometimes of inferiority complexes about the supposedly "more advanced" lifestyles there. As discussed above, Japanese husbands of Filipinas are characteristically seen worldwide as oppressors of "poor Asian" women and far different from "liberated Westerners." Despite achieving the world's second largest economy, in their own and international discourses the majority of Japanese continue to suffer from the sense that they are not quite "modern" in the global hierarchy. Interestingly, these ambivalent positionalities engender soft spots in Filipina–Japanese interactions especially when the men are faced with their wives' "American" behaviors. The location of "America" may be reversed in the homes of Filipina–Japanese families in Japan.

Sitting next to her husband Wada, Amy proudly told me, "Filipino culture is not quite 100 percent, but 90 percent the same as 'American' culture, and romantic relationships are like those in 'America'!" Wada's silence appeared to be dismissive, but also discomforted. Some husbands' dismissive responses are a way to suppress their uneasiness and to tame the power of "America" executed by their wives' behaviors, including the use of English. Similar to other husbands with little competence in English, Oka commented on Filipinos' "strange" English, telling his wife Imee, "If English in America is the real one, Filipinos' English, with that strong trill 'r,' is not." After observing Imee's interactions in English with Americans and Filipinos at her church, Oka felt chagrined by his lack of English ability, and started attending English conversation classes. Imee became his home teacher. This is significant in the two Asian contexts, where being a teacher is considered holding a superior position to be respected. In this way, Filipinas' competence in English, however diverse in actual mastery, allows them to pull themselves out of the "backward South" and reposition themselves in the "advanced West," revealing and displacing the arbitrariness of gender and national hierarchies. When the locations of these men and women in the previously unambiguous gender and national hierarchies begin to crumble and when their homes become hybrid spaces of "traditional Asia" *and* "modern America," their relationships begin to take a different shape. One such example was observed at a family gathering.

In 1996, the Ishikawas held a home party. In the living room, four husbands of Filipinas and a Japanese–Japanese couple, the Hiroses, happened to share the coffee table. The Japanese husbands knew each other through participating in their wives' socializing activities. After they began to feel relaxed enough to chat with the Hiroses, the four husbands started lecturing them. Kawai asked them, "You think Filipinas are submissive (*otonashii*), don't you?" While the puzzled Hiroses were trying to come up with a proper response, Kawai continued, "Nah, in the Philippines, women are stronger than men, so . . ." Taking over, Terami—Rita's husband mentioned earlier—chimed in, "Not only that! Men and women are like this . . . ," moving his hands horizontally at eye level to symbolize the equality between spouses. He then stated, "and, over there, husband and wife are *pātonā* (partners). Or, they say *asawa* in Tagalog! Long ago, the Philippines was a colony of Spain and America. So, Filipinos have been Westernized. Got it?" Ignoring the Hiroses' perplexity, the four Japanese husbands looked at each other and agreed, "Right?" Nods and grins reflected their mutual affirmation. Meanwhile, the men's wives were overhearing this conversation in the kitchen and giggling to each other about their husbands' struggles to come to terms with their unexpectedly "Western" wives and their destabilized gender-national relations. This unrehearsed yet highly mutual experience is suggestive of the de facto collective influence wives have on their husbands in their personal relationships.

In these quotidian negotiations, these Filipinas implicitly and explicitly mobilize a rhetorical "America," English, and their historical "Western" heritage to transform their relations with their Japanese husbands into those that are egalitarian and understanding, and therefore "modern." As the case of Rita demonstrates, these newly created hybrid spaces allow Filipinas to flexibly shift from one self to another, moving between *inter*-national affiliations and gendered positions and between being "single mothers" from the "poor South"—who need the acceptance and financial support of their husbands from the "rich North"—and "Westernized" egalitarianists who speak English, and are boisterous *kamakake* challengers. In the process, the women become mediators of modernity for themselves and for their husbands. These expedient positionings of modern self, however, do not always guarantee happier, more egalitarian spousal relationships.

Divorce, Violence, and the Remaking of Life Abroad

Aligned with the view of Filipina–Japanese marriages as problematic, international audiences at my talks on these unions are extremely curious about "their" high divorce rates. Divorces among these couples began rising after the Ministry of Justice announced in 1996 that it would allow the foreign custodians of Japanese children to stay on in Japan. In 2002, the ratio of the number of divorces (n = 3,133) divided by that of marriages (n = 7,630) showed that 41.1 percent of Filipina–Japanese marriages dissolved, a 20.8 percent rise from 1995.[28] Yet, compared with the 38.1 percent divorce ratio among Japanese–Japanese marriages in 2002 (274,584 divorces over 721,452 marriages), Filipina–Japanese marriages are not so peculiar. Some of my interlocutors nonetheless insist that "Filipinas'" high divorce rates are unique. Although these observers are perhaps sympathetic, such an insistence on ethnic-national difference tends to reinforce stereotypes of the women's victim status. It may also ironically normalize their marital dissolutions because such stereotypes are confused with reality, leading to further breakups. Although reasons for divorces are complicated, many of the broken unions I came to know suggest that Japanese husbands' stereotypical views of Filipinas as being "young and innocent" women of "poor" origin and other defiled images such as being "prostitutes" exacerbated already strained relationships. Some husbands appeared to overcome their own negative self-image of "unable-to-marry" men compounded by their ubiquitous oppressor stereotype by using violence against their wives. Instead of pursuing such pernicious representations, in this section, I therefore discuss the case of Vicki's divorce and the meanings of a tiny triumph she attained as a Filipina divorcee in Japan.

In the early 1990s, Vicki, a former talent in her early thirties in 2001, married Aoki, a member of the police force. Soon after the wedding, Aoki banned Vicki from going out of the house without him. Meanwhile, she gave birth to two children, Erika and Masaya. After seven years of social isolation, in the spring of 2000, a heavily stressed Vicki took her children to the Philippines for a vacation. She returned home to Japan only to learn that Aoki was having an affair with another Filipina, whom he had brought home on a number of occasions. In order to reestablish her economic base in case of divorce, Vicki started working at a nightclub—one of the few job options with relatively good pay available to new Asian immigrant women in Japan.

For Erika's birthday in December 2000, Vicki took the children to an amusement park and a man, Iijima, whom she met at the club, drove them. Upon their return (without Iijima), Aoki was very drunk and interrogated Vicki about whether or not she was flirting with men she met at the bar, although Aoki himself was keeping a lover. Infuriated, Vicki challenged him with the lie that she was indeed having an affair with Iijima. An enraged Aoki then used Vicki's cellular phone to call Iijima and demanded that they meet. Meanwhile, Aoki beat Vicki and leaned over her with a kitchen knife poised against her throat. Vicki managed to escape and saw Iijima coming in her direction. She was so brutally bruised that at first he could not even recognize her. Soon after this incident Vicki left Aoki, and Iijima helped her settle in a new apartment. Physically and psychologically hurt, economically uncertain, and hearing Erika telling Masaya to call Iijima "papa" for his kindness, Vicki became emotionally vulnerable. Knowing about his feelings for her, she soon let Iijima cohabit in the apartment and they started an intimate relationship. However, she also began to feel used when Iijima started asking her to perform "wifely" tasks for him such as making lunches. Before long, Vicki asked him to leave the apartment.

When Vicki's marriage with Aoki began to fall apart in the spring of 2000, her Filipina neighbor phoned me to ask for advice on Vicki's behalf, and I explained that Vicki could seek assistance from public offices and non-governmental organizations in Japan.[29] Sometime after this, Vicki dropped her reliance on men. She moved with her children to a public housing unit. Then, Aoki began to loiter around it, and he gave his children stuffed dolls and a television in which he had installed wire-tapping bugs. He also "visited" them in the middle of the night, jangling the doorknob and insisting that he wanted to meet his children. At last, Vicki decided to contact the police to report his criminal acts. Despite Vicki's report, Aoki has never been arrested, fined, dismissed, or even suspended from the police force.[30] Meanwhile, she learned about free legal assistance available at the city hall to file a demand for child compensation at the regional court.

In January 2002, Vicki contacted Aoki's supervisor at the police force when she learned that Aoki was getting married to his Filipina lover, leaving Vicki and the children without alimony or child support. Vicki sought the supervisor's help in preventing Aoki's marriage until he agreed to her claims for assistance. Throughout the talk Vicki (and I) had with the supervisor, however, he sided with Aoki and criticized Vicki for working as a hostess, overstaying her visa prior to her marriage, and telling Aoki, a man, that she was flirting with other men even if it was a lie. The supervisor drew on many of the stereotypes of Filipinas in Japan that have made them "bad," in his attempt to defend his subordinate and a fellow man and national from the "bad girl" Vicki, whom he felt "undeserving" of compensation. In response to my question about the law for the prevention of spousal violence and the protection of victims, which had just been instituted in October 2001, this man showed no intention of enforcing it.

The next day, Vicki headed for the regional court to file a claim for child support. She won this case. With this victory, she also enabled herself to submit an income report with the welfare office. The welfare office had been making life even harder for Vicki, threatening to reduce her livelihood compensation if she did not find the means to earn some of her living costs on her own.[31] In Vicki's case, ruthless social circumstances have tended to keep her within an isolated home or to push her to fall back on to the help of an informal network mostly among Filipinos, whose knowledge is often limited or stereotyped. However, she was truly enraged by the ways Aoki, his supervisor, and some other Japanese treated her and so explored the options she could find within Japan's legal and welfare schemes.

After winning her legal case, in tears Vicki told me, "Aoki wouldn't have even imagined that *I, a Filipina*, could do something like this!" This may sound insignificant compared to her struggles. However, a majority of Filipinos in Japan are reluctant to visit even the Philippine Embassy for consultation due to the staff's arrogant elite behavior and language—ironically highly technical English full of jargon and acronyms—used toward working-class citizens and especially the (former) entertainers or assumed "prostitutes." To resolve complicated cases many ordinary Filipinos are even more scared to ask for help from the Japanese bureaucracy because of the language barrier and because of their images of the Japanese as racists, traitors, and *kempeitai* (military police during Japan's occupation, 1941–1945). Therefore, Vicki's triumph might seem small when viewed against the forces trying to intimidate her, but it was very meaningful to her personal life. She overcame all these anxieties while stating what a Filipina former "illegal entertainer" and divorcee in a foreign land can do and criticizing the demeaning images and treatment of Filipinas thereof. Her tears in this context indicate not the dismay, misery, and rage of a victim, but the victory and accomplishment of a woman, celebrating her new self and ability to enact her knowledge about Japanese system, whatever valences of power and control still surround her.

Conclusion

Some postcolonial scholarship suggests that people seeking modern social and material relations or identities are, at the same time, subjected to other forces that stem from freedom itself. Then, what the Filipinas described here have achieved may be interpreted, paradoxically, as legitimizing the power and desirability of the North and of America, the "land of freedom and opportunity," while simultaneously reinforcing gendered, classed, and nationalized images of the women such that they would do anything to "eke out" a living in order to improve their "impoverished" lives. Although we cannot dismiss such ironical consequences, in this chapter I have suggested the importance of heeding under-explored issues, specifically the links between agency and mundane processes that generate personal meanings in pursuing, even at great cost, alternative modern lifestyles. As I have shown, everyday lives indeed constitute sites of intense negotiations where people, especially those on the underside of power, maneuver a diverse range of tactics to transform current social and economic relations and forge new subjectivities.

Filipinas in Japan today are not simply there as migrant workers at bars or as *hanayome* assuming the lowest-ranked membership in rural households. A significant majority of Filipina residents in Japan are (were) married to Japanese men and are reworking their intimate lives in new local, national, and transnational contexts. These Filipina migrants and residents did not leave the Philippines solely because of dire financial necessity or filial duty, as the prevailing allegories suggest. Because scholarship has based its analyses only on these structural explanations, and does not investigate women's individual circumstances, it has constructed Filipinas as "bad" transgressors from ideologically "proper" women's roles, places, and sexual morality and as victims of the oppressive and nonnegotiable "cultures" of the "backward" South and Japan.

However, my ethnography has shown that besides such family-based needs, sentiments, and identities, many Filipinas—especially younger women such as Marissa and Julie—pursue multiple desires and situate themselves within modern symbols and ways of life and refashion new subjectivities. In the process, Julie and the wives at the home party boldly and subtly disrupt their representations as "young, innocent, and submissive" women from the "traditional South." Rita even turned stories of Filipina talents as narrated "pauper entertainers" and "single mothers" into a *kamakake* tactic to assure her husband's understanding and bring him into her social world. Exploring newly acquired knowledge, the "former illegal and promiscuous entertainer" Vicki was able to combat violence and injustice and, through winning an unexpected legal claim, reposition herself and her children in a foreign land. In so doing, these women express their material, gender, and national parity with people of the "North," but with a Filipino spin.

The most significant theoretical challenge to the existing order and location of power as previously assumed are most intensively played out in syncretic arenas opened up between two Asian peoples with different histories. The Filipinas explore their multi-sighted dispositions, affiliations, and knowledge, based on the hybrid origins of Philippine culture and history and of their locations in the global hierarchies, as well as on Japanese men's ambivalent identities. These have engendered dynamic sites of negotiation between spouses in which the women stir up classed and nationalized hierarchies of signification such as the English language and Western colonial history. In the process of searching for alternative modern lifestyles and identities, the Filipinas and their husbands have intervened into the rigid analytical categories of North–South, East–West, women–men, and other binary divisions. Such multidimensional positionings of one's identity do not create utopian relations. Nonetheless, this emergent space evokes many more manifold imaginaries about each other and enables the women to tactically shift from one identity to another. Meanwhile, no matter how the structural discipline subjects them, in their everyday lives these Filipinas themselves will not cease— as their ultimate moments both of resistance to such domination and of achieving alternative modern subjectivities—to exert their own power to think, feel, care, desire, and act in creating personal meanings and in nurturing affective relationships.

Notes

Special thanks go to Vicki who allowed me to take part in her courageous endeavors to better the situations of distressed Filipinas in Japan. I also appreciate all the other Filipinas and their Japanese husbands who shared their everyday lives, laughter, and tears with me. I am grateful to Laura Miller and Jan Bardsley for their never-ending "bad" intellectual stimuli, colleagueship, and "catty" support. Sarah Frederick and Kawahashi Noriko also offered valuable comments.

1. Here and in my fieldwork, I call Filipinas by their first names and Japanese by their family names, all of which are pseudonyms. All the quotes used in this chapter were obtained through interviews and casual conversations with my informants in Japanese, Tagalog, English, or a combination of these languages. Ungrammatical renderings are not intended to discount their linguistic skills but to deliver the tone of their original utterances, as even native speakers rarely talk flawlessly in colloquial conversation.

2. For details, see Nobue Suzuki, "Between Two Shores," *Women's Studies International Forum* 23, no. 4 (2000a): 431–444.

3. See for example, Karen Kelsky, *Women on the Verge* (Durham, NC: Duke University Press, 2001).
4. See Catherine Ceniza Choy, *Empire of Care: Nursing and Migration in Filipino American History* (Durham, NC: Duke University Press, 2003).
5. Dilip Gaonkar, "On Alternative Modernities," *Public Culture* 11, no. 1 (1999): 1–18.
6. Fenella Cannell, *Power and Intimacy in the Christian Philippines* (Cambridge: Cambridge University Press, 1999).
7. The number of Filipino spouses is surmised based on people who hold "spouse-or-child" (n = 45,510), "permanent" (n = 32,760), and "long-term" (n = 18,246) visas in 2002. About 10 percent of the "spouse-or-child" visa holders are children. Many long-term visa holders are former spouses and the custodians of Japanese children. See Ministry of Justice, *Annual Report of Statistics on Legal Migrants* (Tokyo: Ōkurashō, 2003). Of the total Filipino–Japanese marriages, 1 percent consists of Filipino men–Japanese women unions. A majority of Filipina–Japanese couples throughout the 1990s, and 40.6 percent in 2002, have lived in Tokyo and three surrounding prefectures followed by the Tōkai area consisting of Aichi, Shizuoka, and Gifu prefectures (16.6 percent of registered Filipinos). See Ministry of Justice, *Annual Report* and Ministry of Health and Welfare, *Vital Statistics* (Tokyo: Ōkurashō, 2003).
8. Commission on Filipinos Overseas, "Number of Registered Filipino Fianc(é)es/Spouses of Foreign Nationals by Major Country of Destination: 1989–2001." Online at <http://www.cfo. gov.ph/>.
9. See Hagi Kazuaki, "Yokatta, yokatta . . . dake dewa," *Asahi Shimbun*, October 25, 1987, 20–21.
10. See Mike Douglass, "The Singularities of International Migration of Women to Japan," in *Japan and Global Migration*, ed., Mike Douglass and Glenda Roberts (London: Routledge, 2000), 91–120; Wolfgang Herbert, *Foreign Workers and Law Enforcement in Japan* (London: Kegan Paul International, 1996), 99–120; Muriel Jolivet, *Japan: The Childless Society?* (London: Routledge, 1997), 147–161; Gavin McCormack, *The Emptiness of Japanese Affluence* (Armonk, NY: M.E. Sharpe, 1996), 178–180; Yoko Sellek, "Female Foreign Migrant Workers in Japan," *Japan Forum* 8, no. 2 (1996): 159–175.
11. For a fuller account of the images of Filipinas in Japan from the 1980s, see Nobue Suzuki, "Women Imagined, Women Imaging," *U.S.-Japan Women's Journal English Supplement* 19 (2000b): 142–175.
12. Douglass, "Singularities." In March 2005, the Japanese government began tightening restrictions on the entry of workers with entertainer visas. Observers are skeptical of their seeming feminist rhetoric of wanting to "protect women from being trafficked." This move must be critically read within the context of heightened nationalism, a diluted national "mono-ethnic" body, and increasing welfare costs for (foreign) single mothers.
13. Herbert, *Foreign Workers* 29.
14. For details about talents, see Laura Miller, "Crossing Ethnolinguistic Boundaries," in *Asian Popular Culture*, ed. John Lent (New York: Westview Press), 162–173.
15. McCormack, *Emptiness*; Jolivet, *Japan*.
16. Douglass, "Singularities." For more complicated life trajectories of eldest daughters and others, see Nobue Suzuki, "Transgressing 'Victims,'" *Critical Asian Studies* 35, no. 3 (2003): 399–420; Nobue Suzuki, "Tripartite Desires" in *Cross-Border Marriages: Gender and Mobility in Transnational Asia*, ed. Nicole Constable (Philadelphia: University of Pennsylvania Press, 2005), 124–144.

17. Tomoko Nakamatsu, "International Marriage through Introduction Agencies," in *Wife or Worker?: Asian Women and Migration*, ed. Nicola Piper and Mina Roces (Lanham, MD: Rowman and Littlefield, 2003), 181–202; *Immigrant America: A Portrait*, ed. Alejandro Portes and Ruben G. Rumbaut (Berkeley: University of California Press, 1996).

18. Sellek, "Female Foreign Migrant Workers" 160, 168, 172, emphases added.

19. Douglass, "Singularities"; Herbert, *Foreign Workers*; Jolivet, *Japan*, 157; McCormack, *Emptiness*, 179.

20. Nobue Suzuki, "Inside the Home," *Women's Studies* 33, no. 4 (2004): 481–506.

21. Philippine Women's League of Japan, "Pinays in Japan." <http://japan.co.jp/~ystakei/~pwl1.html>.

22. Dinah Jowan, "Interview: Watashi no Japayuki taiken," *Bessatsu Takarajima* No. 54, *Japayuki-san Monogatari* (Tokyo: JICC, 1986), 113–121.

23. Upper-class Filipinos had long enjoyed material, cultural, and symbolic wealth far beyond today's petit bourgeois Japanese, even before the penetration of Japan's capitalism.

24. Cannell, *Power*.

25. See Nakamatsu, "International Marriage" and Suzuki, "Transgressing" for further critiques of the view of the "bride trade" involving international marriages arranged by introduction agencies in Japan.

26. Nobue Suzuki, "Gendered Surveillance and Sexual Violence in Filipina Pre-Migration Experiences to Japan," in *Gender Politics in the Asia Pacific Region*, ed. Brenda Yeoh, Peggy Teo, and Shirlena Huang (London: Routledge, 2002), 99–119. Annulment of marriage is a possibility, but many Filipinos simply leave their marriages without complicated and costly legal procedures.

27. Here, I am specifically referring to the *Nihonjinron* (theories of Japanese uniqueness) upon which such a contrast is built.

28. Ministry of Health and Welfare, *Vital Statistics* (Tokyo: Ōkurashō, 1996). Unlike Japanese intra-ethnic divorces, there are no statistics on divorce cases per 1,000 population among the non-Japanese population.

29. Before proceeding, I should clarify the roles I played in this case. As a policy, I refused to make any decisions for anyone, including Vicki. Instead, I explained her legal rights and terminology and provided the names of government and non-governmental organizations. The only time I directly intervened into this process was when Vicki was frightened that Aoki might set their house on fire. Another time was her visit to Aoki's supervisor described further. Despite whatever I may have done for her, it was Vicki who made use of the advice and information she acquired from others and me in the remaking of her and her children's lives.

30. Although she kept evidence of Aoki's stalking, Vicki did not want to push the criminal case too hard because of fear that her children might be affected by the publicity of their father's unlawful activities.

31. Governmental responses to these foreign women in distress are inconsistent. Some were unnecessarily supported even though they insisted that they wanted a job not welfare funds. Others, like Vicki, were pressed by welfare officers to find work amid severe difficulties without themselves providing much assistance.

11

Sex with Nation: The OK (Bad) Girls Cabaret

Katherine Mezur

Welcome to "Suffocating World!"
Because women lack physical strength, they dream of being skunks,
—in an elevator, a lonely alley, walking alone at night—
A skunk for any situation. . . .
I don't want to get dirty, so don't walk on my skirt.
Cover yourself completely. Come to me.

Norico Sunayama[1]

Feminist body art crosses over subjects and objects of cultural
production in a chiasmic interweaving of self and other that
highlights the circuits of desire at play among them.

Amelia Jones[2]

When her skirts ripple, the room becomes a tidal wave of red, a red sea of cloth. The gallery is Norico's dress. If Norico pulls at her skirts, would the walls come down? Or are we drowning in her red dress? The red skirts flow away from her small gesturing figure. She is the vortex and the siren, sitting on a twelve-foot high pedestal. Norico embodies the room and becomes the dress consuming the room and all beneath. One enters the "exhibit" by crawling under her skirts. Billowing and swimming your way through the red skirts, you arrive at Norico's throne, a padded toilet seat on a raised platform. Under her skirts that form a tent beneath this throne, everyone is bathed in the deep red light filtering through the crimson fabric. Norico invites spectators to take a flashlight and look up her dress to her seated bottom. Wearing white little girl underwear, she moves on her seat, rocking back and forth, lifting up and plumping down again, making billows of her dress swell, surge, and roll away. Her thighs press against the seat. Everyone underneath glances up, then away, caught in the act of looking, they are the captured audience, bowing beneath the bad girl goddess's skirts.

This is "Suffocating World," an installation project first designed and performed by Sunayama Norico in 1996. Norico installs "herself" in a giant red dress that fills the entire art gallery, a wall to wall dress. Some spectators barely enter the dress before quickly turning around, embarrassed to find themselves under her skirts, beneath her bottom. Some wander about, lost and confused in her "suffocating world."

Norico navigates her dress, addressing the way she feels caught or contained in prescribed appearances. She says she took her mother's admonition that her short skirts were indecent and blew up that little girl skirt into this exhibition. Now that dress is the suffocating space of the spectators. Norico takes the suffocation idea one step further by referring to a skunk in her program. Recalling how the skunk's fear produces its odor and chases people away, Norico burns lavender incense beneath her skirt. She invites and repels. She fabricates her own dress-code and creates her own peep show, flashing the stealth voyeurs who cop panty-shots with hidden cell phone cameras. Her red dress can be read as an inversion of the Japanese flag: the rising sun circle has bled to the edges, a sunset or bleeding stain, with Norico dead center, as the new neo-moon goddess. The red is the border and at the center is Norico's white bottom framed by a toilet seat. Norico's deliberate confusion of private and public space, individual and nation, invitation and tease are brilliant bad girl acts.

In the following essay, I examine Norico's performance art and installation as bad girl strategies for gaining a kind of agency that is specific to the bad girl culture she creates. I also suggest how her girl culture is simultaneously produced within Western globalized commodity culture, but resists that hegemony. I consider how globalization's processes of assimilation, appropriation, and homogenization are transformed through Norico's girl/woman frame of ero-porn acts and their local refractions and differences. In one of her performances, which I analyze here, Norico seems to dare girl culture into badness. That is, by masquerading through the stereotypical images of the status quo bad girl, she forces the spectators and herself to glimpse a futuristic polymorphic girl culture that could excite and mobilize fluid national/cultural orders.[3] Norico disorders hierarchies: her bad girl acts are of that different order. Norico's choices of choreography, costuming, music, and visual display, like her giant "suffocating" dress, and "exhibition" of her bottom-beneath, indicate transgression on one level and visionary play on another. Norico's bad girls' girl culture enjoys an agency that may appear transitory and limited to her experimental art world. However, her "bad girl" girl culture persists and reverberates, re-enforcing a spectacular Japanese girl culture.

Certainly the ongoing traffic in Asian female bodies, throughout Asia and beyond, speaks to a global connection of women as national products, and transport bodies of culture. Mainstream globalization runs on two tracks: one for the consumers or the mobile global middle and upper classes of nations and another for the consumed, the lower working classes who are sometimes mobile and sometimes fixed. Norico's performances offer an entirely alternate route that spins off of the particularities of gender, sex, and sexualities. A wild ride, her works disrupt those inter-relationships in different cultures and disorder the modern, Westernized urban market/performance spaces. Norico has performed her acts in gay clubs in

London, gay-on-some-nights clubs in Kyoto, and as post-show cabarets in the United States. My reading of Norico's performances could be considered in terms of a queer global arena, however, I am wary of equating her work to a drag king or transvestite performance in Sydney or San Francisco. As Dennis Altman emphasizes, globalization may appear to "transform local regimes of sexuality and gender,"[4] but we should be careful not to conflate diverse sexual identities with those based in the Euro-American urban spaces.[5]

Girl Nomads

Placing Norico's work and that of Japanese women artists like her in a larger framework, I consider how these performances are products of the ever-present national coercion for globalization that is constantly intercepted with counter surges of exclusive Japanese-ness. Between these forces, young women and girls play out complex and shifting mutations of the Japanese nation's global agendas. In the arts, women have created radical body/media performances, video art, and installation performances that defy any one category of art making. They play with high art aesthetics, erotic pop media, fashion, and commodity culture. This close-up performance analysis reveals not only the cracks and decay in Japan's global urgency and the underbelly of the nation-state system, but also hints at a new cultural address from the dynamic position of a nomadic local girl who manages to traverse and perform (in) her own extra-global terrain.

Globalization has become a practice, an act of everyday life. But as Trinh T. Minh-ha observes, it may be a term that has been emptied out and no longer "means" but echoes and reverberates perhaps making itself look and sound larger than it really is.[6] It is in this world of global echoes that I wish to interject "girl genders" in performance, in this case specifically the gender performances of Japanese women performance artists, who deliberately play with global hegemonic images and acts. Their genders multiply and skew the globalization pattern of oppositional and exclusively male and female categories and the homogenization of erotic attraction. In particular this examination focuses on the image of Norico's sexed-up-and-over girl in the global club space, and how she disrupts the global circuit of female bodies.

Given the long history of colonial hegemonies and forced migrations, globalization has become more complex and more disturbing with technological advances. As Dennis Altman argues in *Global Sex*:

> In a globalizing world, with technologies quite different from those of pre-World War I expansion, time and space themselves take on different meanings, and no aspect of life remains untouched by global forces. Yet most people still operate within particular local spaces, and there is constant tension between the local and the global.[7]

It is this tension between local and global that foregrounds the incredible differences and particularities that globalization ignores or attempts to homogenize for the global marketplace. Altman examines how sexuality is one arena in which global strategies have met local resistances that have, in turn, created new forms that confound the homogenization trend of global capitalism.[8]

In this chapter, I want to show how the experimental performances that deal with sexuality and eroticism reveal multiple and counter responses to global marketeering and products. Norico's bad girl performances discussed here, then, are occasions for turning the close-up lens on global leaders, global sex trade, and global violence. The performance performs the shifting of global and local acts in the context of erotic entertainment. Norico takes the opportunity to use her marginal vantage point to mimic the global center. In this exposure of her "bodies," exhibiting cultural and global processing, the stakes are very high: this is where cultural change happens or dies. In this case, Norico plays with evolving gender and sexual role types, thrusting her own, local, exaggerations of racial and sexual acts into the faces of those who proscribe global homogenization and stabilization. Norico's bad girl acts aim at a differentiating global: she performs a global dissemination of differences and possibilities.

Framing Local Girls

Whereas significant progress has been made in changing attitudes toward women in the public sectors, many women and girls in twenty-first century Japan still occupy marginal positions in all states of life and occupations, including the arts. However, for the last fifty years in the arts, this border area has gradually become a space of frenetic, exciting, volatile, and highly influential creativity as well as a source of inspiration for commercial and global commodification. As argued earlier, the Western-based concept of agency needs to be reinterpreted and reconsidered in the Japanese girl cultural space, where power or authority may not be named or credited in any way, but is still there. In some ways, the energy and brilliant visual display of J-girl culture has made it highly commodifiable in foreign urban cultures. By blowing up commercial J-pop products, already hybrid, and mixing these with the global and techno sex/body markets, women artists are pushing and disturbing those very meanings of nation, global, and culture. Their outrageous, rebellious, and neo-postcolonial acts are woven into and driven by a saturated commercial culture, which they take on with skillful vengeance. The mix of gender roles, erotics, sexualities with Japan-pop, post-Americana, and *shōjo* (girl) culture is fascinating and abject. These performers make use of the magnetic draw that they can kindle in viewers with their outrageous acts: using their moments on the stage to create fissures with the Giant of Japan-nation-state, and any hegemonic cultures. Theirs is a deliberate shakedown of the ubiquitous "Asian woman" idol, a ride through the stereotypes dressed, stripped, and gender-altered. These performances demonstrate how bad girl culture could be a new pattern or model of an anarchic globalization.

In the mid-90s, the revenge of the cute little girls produced the extra-ordinary *kogyaru* (little girl) culture that was like a "blow-up" or super-sized, even grotesque, version of the enduring and more endearing *shōjo*. The little, cute girl image, repeatedly force-fed to girls and even women, is held up as an ideal of the feminine.

It is little wonder that its ubiquity prompts artists like Norico to take over those girl images and acts, embrace them, and then manipulate them to their own advantage and then some. In Western feminist theory, agency or the power to shape and have control over one's body, self-image, and identity is essential to empowerment in daily life. Certainly, this was also one theme in Japanese modernist feminism. In Japan, as in Western countries, women still do not have carte blanche when it comes to controlling their own bodies. In a sense, this takeover, or I argue, this mutation of the little girl body and her identity, however, open up a whole new territory of abject agency.

To examine these performances, it is important to think performatively and witness these acts in the moment of the "live" encounter: the charged, sensual, and violent woman meeting the adorable, precious little girl. This confrontation certainly suggests many possibilities for multiple identities (doubled, mixed, ambiguous, ghosted) and their own form of agency with those new forms of selves. Also it helps to think of how power and agency are derived from a culture-specific hierarchy and its oppositional system of gender and sexualities. In contemporary Japan, in its present male hegemonic system, occupying the woman-identified position and identity will not have a great effect. In contrast, young girls and women, recognize possible and productive agency from within popular culture (which is the center and the goods of Japan's globalization). This productive, even radical, agency is not over and above male power, but rather beside, behind, beyond, inside and through that male-powered world. The OK Girls of this chapter, do not destroy the little girl. Instead, their agency works through saturation. Girl acts are drenched with longing, sweetness, desire, and anger. They channel an enormous backup of energy, creativity, anguish, anxiety, erotic desire, anger, violence, and fun or exuberance into new works of art and performance. But creations do not happen in a vacuum. These works disrupt the exotic and erotic power-packed global commodification of "female."

Norico's Global Glamour

Sunayama Norico, a choreographer and an installation and performance artist, is one among the new breeds of nomadic local girls. She is nomadic because while she tours and travels with the internationally known performance art group, Dumb Type, and because she performs her own "cabaret" in clubs wherever she is touring as well as at home in Kyoto. She travels but her performances are deeply concerned with her local place, urban Japan. According to a special issue of *Montage* dedicated to Norico, she is, in the magazine's English, the "Gura gura Glamour Inferno!" who is "Worked up of constricted, Fascination of Valley between."[9] The articles describe the range of her "glamour activities": her work with Dumb Type, the famed high-tech performance group collective with a progressive political agenda, her club acts, live installation art, and creation of the "OK Girls." Sunayama Norico has been performing in contemporary performance and dance since the early 80s. Based in Yokohama and Tokyo, she danced with the dance group, Kurosawa Mika and Dancers. Kurosawa, who is an exceptional woman choreographer/performance artist who draws on Butoh, modern dance, and performance art, lived in New York in the early 80s. Kurosawa's body-focused art is greatly influenced by American performance art and contemporary dance theatre Norico and Kurosawa represent a small network of women artists who have created works and groups that were either women-only or dominated by women performers, whose content focused on the female body, sexuality, and erotics. While most of these women have said they have done women-only work, none of them articulate that they are feminist. Instead, they enjoy their rebel, avant-garde, or "bad girl" status that keeps them safe from being labeled "feminist," a title as problematic in Japan as it is in the United States, as I discuss later in this chapter.

In 1990, Norico was invited by Dumb Type's leader Furuhashi Teiji to join the Kyoto-based performance collective, which he had formed in 1984, as one of its central performers and choreographers. Furuhashi, who died of AIDS complications in 1995, was an activist artist. His sociopolitical direction of Dumb Type facilitated Norico's move into her own activist performance work. For example in Dumb Type's "S/N," one theme focused on sexual identities. To pulsating electronic music, giant projections of "medical" photographs of bare torsos (company members') without heads, flash in sequence across the stage. We cannot see who is "male" or who is "female." The visuals disturb the established binary opposition of male and female and heterosexuality. Across the top of the projections in a loft-like stage space, silhouetted figures, run, touch, and jump off the stage, disappearing in a flash of strobe light. At the same time, a printed text runs across the wall, asking questions about pleasure, sex, desire, and power. Dumb Type's work was most often abstract, making its political punch through the repetition of images and text. Norico and the other women shaded that abstraction with more erotic and emotional images and activities such as Norico's choreography and performance of herself as a naked doll being walked by a group of white-clad doctors in "OR," or in the S/N finale where performer Bubu pulls a long string of international mini-flags out of her vagina.

New artistic directions emerged when Norico met Bubu, another Dumb Type performer, who was also a sex worker activist, and Mami, a lead woman performer and writer in the media-pop-performance collective, Kyupi-Kyupi. In 1995, Norico formed the "Ok Girls," with these women and other Dumb Type women, who along with AIDS activists, took up the rights of women and erotic/sex workers. Among their politically-charged works were the celebratory, fun "shocking" drag king shows in clubs, theatres, and museums.

Norico's OK Girls are special "bad girls" who work in the subversive current of women artists who create the rifts in the masculinist establishment in the performing arts. All these women share a keen sense of gender transformativity through the use of excess, drag, and saturated obā (over) eroticism in works that share an underlying politics: a hidden bad girl agenda of overt sensuality and pleasure. Norico calls her agenda, "a positive sex evolution . . . since there are no contemporary heroines, we are still searching for the mythic . . ."[10] According to *Montage*, Norico's most famous themes are: "Safer Sex Revolution, Positive Sex Evolution, Masturbation Invention, Super Fabulous Glamour Activities." Among her subtle bad girl acts is her preference for being known simply as "Norico."

She drops her family name, which usually comes first and foremost in Japanese naming practice. Without a family name, one has no class or citizenship in Japan. She also writes Norico without Chinese characters, which are usually chosen especially for the individual child. She uses *katakana*, a writing system used to delineate foreign words and emphasize non-foreign words, thus making her own name "foreign" or imbued with an outsider status. Norico also changes the final "ko" (meaning child) to "co." The child "ko" character that is frequently attached to girls' names renders them indisputably female and associates the feminine with the child-like. Norico dispenses with all those associations. She also asserts that her "co" is the abbreviation of "Co.," for "Company." She is her own, one-person company.[11] Norico's transformation of her name is a bad girl paradigm: within the system, she changes what is possible to change—what is visible and subtly disturbing. Her performances use a different strategy: outside the (inside) system bad girls let go.

Norico continues to create and perform with Dumb Type[12] and whenever possible, she performs her OK Girl Cabaret after Dumb Type's main stage shows, usually in the venue's lobby or nearby club. Norico's OK Girls cabaret club acts are fun and provocative mixtures of karaoke songs and drag-like acts choreographed by Norico and the other performers. In the process of investigating Dumb Type and Norico's performances, I was able to see them perform in the United States, Canada, and Japan. While this is not a global sampling, I have witnessed the processing of these works over time and in different cultural spaces. Over time and space, Norico has both deepened and expanded her national-cultural critique. In her most recent work, *How to use the weapons*, she expands her OK Girls' agenda: in a world of male global leadership, in which money and weapons dominate, the OK Girls perform transformative erotic and abject acts that render male weapons obsolete. All is OK!: OK (Bad) Girls rock and entertain.[13]

Bad Girl "fun" and Abject Agency

"Ero-porn" acts according to Norico, combine seductive gestures with stereotypic pornographic poses and moves that are performed as a kind of masquerade and "making fun." These acts of fun ero-porn are a part of an exponential boom of girl agency. But what kind of agency is this? Perhaps when Western feminists use the term agency, they most frequently refer to possessing power or authority over one's self. That is, one may have agency or authority over one's own body, appearance, and behavior and that power is considered an unquestioned right. Agency is revered. I would like to argue, however, that an individual inhabits a very specific culture-bound moment/space, and as such, agency itself must be defined within the cultural context of the individual. In this particular world of Japanese performance art, agency is reworked simultaneously through the lens of Japanese culture and "girl" culture. This girl agency is leaky because it morphs. It is inauthentic because it borrows and mimics other cultural "agencies." It is mysterious because it conceals itself, and it is messy because it refuses control and demands disorder. This girl agency is abject—it is a bad girl agency.

The women here operate with this abject agency. Their power and actions are always already beyond the regulated norm. They are the bad girls whose creative agency arises in the rich wastes of social aversion. Performing in that created space of abjection, forced outside a regular system of performance that is dominated and run by men, they enjoy a radical freedom. For example, in "Suffocating World," Norico's inner space, under her bouncing bottom, is an abject space of girl agency: once you are under her skirts, you are in her power. She invites and forces the spectators to look up her skirts. They become both shocked and compliant because the rules are switched. "Girl-ness" in Japan so often means being subjugated to the gaze and touch of men, but in this case, men are "suffocated" by the skirts they look under or try to lift. Norico "girls" the act through its exaggeration and abjection: Norico shocks her visitors with her bare bottom and invitation to look with a flashlight. And in this moment of shock—a shock that is all the more shocking because it defies male rules—they have freedom.

One of the sources of the OK Girls' creative fun is the "stuff" of globalized pop culture. Here notions of agency concern disruption, subversion, and violation. Further, according to Norico, her objective is "fun," a letting go to the point where "everything is OK." Her agency is specifically performed in a realm of play or with a sense of fun that does not focus on material accomplishment, but pleasures for pleasure's sake. The pleasure here, her fun, is also different. It is not aesthetically controlled, but instead, it is chaotic and de-controlled on purpose. Everything is *obā* or over the top, overdone, overwhelming.

The "play" of abject acts makes a fluffy and charged "show" that the women creator/performers thoroughly enjoy. Their agency (everything is "OK") does not aim to define their subjecthood or articulate their identities, or express there inner self-hood, rather they "play" fun over-the-top for their own exhilaration. Their ero-porn acts are messy: transformative, ambiguous, confusing, silly, painful, and childlike.

Bad Girl For-play

The examples considered here raise questions concerning how young Japanese women access agency, which allows them to redefine themselves by changing their bodies-as-objects into bodies-as-subjects. To facilitate their corporeal subjectivity, women artists have developed acts of "excess." For example in Norico's Red Dress installation, the dress itself is an act of excess. A gigantic dress that swallows people, a skirt that becomes the architecture of a gallery, a girl's body that is fabricated to enthrone and fetishize the hidden sexual organs. These artists could be read simply as "bad girls" exaggerating and flaunting the techniques of the masculinist regime of female exploitation. To a certain extent this may be so, but according to Norico her red skirt is not just flaunting but "killing" through the suffocation. Her suffocating skirts kill those underneath. While Norico invites you to enter and look up her skirts, most "entrants" do not stay to play. Are her bad girl acts invitations to transformation, annihilation, or rejection?

I propose a bad girl hypothesis: perhaps the bad girl acts of excess only tease and trouble the hegemonic masculinist regime, but this troubling is in itself exhilarating. Perhaps "bad girl" here is "excess." Perhaps bad girl is powerful, without any translation into or through the masculinist world in its own separate girl culture-nation. Certainly because of global commodification Japanese "bad girl" culture in anime, manga, novels, and performance suggests the curious strength of oppression: girls turn around the rules of girl bodies as the sweet innocent "to be violated" objects by performing the oppression outrageously. Their power-up strategies show leaks in the masculinist dominant system. After Suffocating Dress, Norico's other acts, especially her OK Girls strip and drag king numbers, suggest that her pursuit of agency may be accelerated by these bad girl strategies.

I suggest that Norico's "bad girl"-ness is the creative wedge of resistance to a mainstream culture that assumes she will grow up. This resistance is a kind of twisted agency that allows her to keep working as an artist and to keep marking her work as a woman artist. "Marking" in this case is strategic: raising a multitude of questions that go unanswered, but at least cause uncomfortable reactions. Her extravagant exaggeration, multi-sexual cross-dressing, and forms of "sex tease without touching" mixed with punk, Play-girl acts of display, and karaoke party, and careful choreographic and scenic techniques, speak of an agency that moves strategically. Hers is a resistant agency: at once national and transnational drawing directly on her materiality: body, clothes, fashion, gaze culture, and media. Her girlness and her badness make her a kind of "double agent": seemingly playing along with the boys, but instead her bad girl agency creates a local/global girl culture that baffles both Japanese and Western exploitation of women and girl culture.

Girl, or little girl in this performance context of excess needs to be reconsidered, according to Larissa Hjorth, a scholar of Japanese popular culture and sexualities, who argues that "cute" and "little girl" characters need to be taken beyond the heterosexual framework, in order to "open up possibilities of nonconformist gender roles and sexualities."[14] Hjorth suggests cute characters like Hello Kitty facilitate a different agency that ". . . enables the configuration of new categories of gender and sexual identity."[15]

Further, Norico's bad girl acts may be considered as "radical critique" in what Amelia Jones calls "the rhetoric of the pose." Jones suggests that the woman performance artist "both exposes and makes use of the conventional codes of feminine display to increase her notoriety . . . in the male-dominated art world."[16] Perhaps when Norico takes on these other bodies with breasts and penises she joins other women artists who,

> complicate and subvert the dualistic and simplistic logic of these scenarios of gender difference by which women are consigned to a pose that is understood to be unself-reflexive, passively pinioned at the center of a "male gaze."[17]

Norico also parodies her self and the larger "selves" through her layers of impersonation, which she "feminizes" with the excess of "skirt" and extra breasts.

Norico's dress installation and her "dress" and drag acts expose the male construction of femininity and "opens the female body/self to other desires and identifications."[18] Norico's bad girl acts flip the sex/gender dominance, for pleasure.

I link globalization with gender because I think they form a partnership that complicates the view of globalization as worldwide circulation, distribution, and communication. Here I will focus on the bodies of women as global commodities and the globalization of "female" gender acts as a resource for Norico's hybrid erotic shows. At the same time, Norico and the OK Girl "acts" globalize girl-culture-nation through the seduction and disparagement of hegemonic nation-state-heads. Norico and OK Girls have "sex with the nation-head" on their terms. Her performance simultaneously reflects the system of coercion that enforces girl/woman global-exchange and revels in their subversive erotic play with that system.

Norico's[19] bad girl identity is her drag-king-like persona, Snacky, and her repertoire of OK Girl acts such as karaoke strip tease, a hip hop bump and grind with a cutting political edge. In her March 2003 *What to do with the Weapons* show, her first "act" begins with her wearing a giant spongy afro with Carol Channing eyelashes and lips, in a slick satiny black pants suit with front zipper, and a red shirt with white ruffles. She twists, struts and rants (in foot-high silver boots) a soft soulful James Brown song: "Cold Sweat." The music breaks to a full classical orchestra rendition of Manuel de Falla's "Ritual Fire Dance," the driving and racing work, originally written to depict the romantic gypsy tale of a two lovers who dance a mystic fire dance to rid themselves of a jealous male ghost. Norico suddenly cranks into high gear and begins her "ethnic" strip: She ties a harem-girl scarf across her lower face, her "oriental-global" mask, shimmering with tiny gold coins à la belly-dance style. Norico is un-mapping the globe as she unzips to her gold chiffon pajama top, that just glances the top of her g-string clad bottom and net tights. She writhes and stomps from soul-boy to harem-girl and on into stripper-SM terrorist, whipping to the war-nostalgia tune of: "We'll meet again . . . don't know where, don't know when . . ."[20] Norico faces off to the boy-nation, challenging world leaders to use their "weapons" her way.

My thesis is that the transnational movement of female bodies (live and mediatized) has created a "link" or transfer site between female bodies and culture nations.[21] I am looking at experimental/political performances based on this "performance" of girl as culture nation representation from the "flip-side" or "inside-out-side." That is, Norico and the OK girls, who are sometimes sex workers and club dancers/hostesses, also create and perform counter, outrageous, serious "fun." Norico calls her club acts "underground" or "cult" shows when she performs in Japan, but on tour, she calls them cabaret acts, the touring versions of the home routines.

OK Girls "Cabaret" and Norico's "Snacky"

Stripping through different ethnic disguises, Norico sets up her acts to rip away at the clichés and play with them simultaneously. Norico's "play" is complicated as she does this for fun that has a purpose to "make fun." Norico maintains that her pleasure is first and that is what she always performs: her sex/sensuality or character/persona, flamboyantly displayed as fun: "everything is OK." This "Ok" is already a loaded sign/product of globalization of American "slang", which has its beginning history in the military postwar occupation. Considering we are in a "way-post" era, Ok has become "Japanized," possessing its own feeling and meaning, and refers mainly to an unrestricted occasion, be it verbal or corporeal. For Noriko it means she can do anything she wants with her body. OK frames her acts as outside of obligations and oppression. As Norico suggests, "OK Girls means everything is OK."

For the opening section for her main cabaret act, Norico opens with a version of her male persona, "Snacky." Snacky is what Norico calls her cult name, which she created from a reworking of "Sunayama" to the endearing "Suna-chan" to the further abbreviation of "snachan" to "Snacky." Norico suggests this also alludes to the English slang "snatch" and "snack." In this role, she first appears as a boy-like "Elvis-impersonation" with Elton John glasses. She reenters the stage disguising her "Snacky" with an Afro wig, a wild patterned shirt and bell-bottom pants, and dark sunglasses. Norico's drag and "strip" act moves back and forth between dressing and undressing, curiously shifting genders and identities. Her bad girl acts deliberately paste on and peel off possible selves, possible identities. For example, after a certain amount of bumping and grinding and singing to her own voice with a giant microphone on a tiny bar-stage backed with floor to ceiling mirrors so we can catch the act from front and behind, she begins to peel off the shirt and pants and reveal her under-costume of "harem girl." Still dancing and jiving on punk rock songs, she eventually goes from Afro boyish to Middle Eastern maiden who is ready for it. She pulls off her shirt to get to her golden outfit of bra and flounce, tiny pants/g-string and black fishnet stockings, and a little black lacey see-through "veil" over her nose and mouth. Little gold coins flicker at the end of the lace, swaying over her mouth that sings of her aching, longing love. This is not love or desire for a specific person, but for pleasure, especially relaxed pleasure. At one point on the video and several times in a real show, Norico crouches down and opens her legs to beckon, to touch, as she caresses the space/place of her vaginal opening that is carefully layered by stockings and g-string.

Bad Girl Nation?

Norico's club, cabaret, and cult persona acts, clearly target something about "Japan" and the production of erotic bodies and their acts in and out of Japan (in a variety of media). Her trans-gender acts are more complicated because she criss-crosses gender and ethnic stereotypes. Her aim is also transnational: She hangs up many of the current "heads" of state, from Japan, the United States, United Kingdom, Korea, and Israel, clearly aiming her ethnic drag at multinational targets. In this section, I focus on one of her performances in which Norico's "girl" contests and confounds this identification with her "global" sexual entertainments/cabaret club acts. While she says she is having fun, what she does flies into the face of the stereotype Japan Girl. Norico remaps a Girl Culture-(Nation) that makes strange or uncanny the trans-national female erotic acts that have become "global" to those within reach or exposure to media-driven visual sex culture based on Western media. In a larger sphere, Norico's acts dump "nation-state" for "girl-nation" because her acts mock the pure Japanese nation-state (a masculinist hegemony) but are exemplary Japanese Girl culture acts. Her acts are "made" in Norico's Girl Japan for local and global export and, in that circulation of touring and returning, become "transnational" commodities. Noriko and other Ok Girls radically mimic a "global (male) consumer" culture's iconography or representations of strip, porn, and erotic titillation which transgress the Japan-Nation/culture and smashes the global glittering circuitry in order to paradoxically reveal its fabrication and destroy the lie of natural or original cultural identity. Norico's fun has to do with trashing the "girl" and recycling girl and global "oriental" female in her own "choice" commodity. Norico may not have agency over the "global" roles, but she performs her mutations of those "other" objects/roles. She "subjects" or girl-inculturates these objects-of-desire giving herself and the audience, momentary pleasure, fun, in her radical exploitation of the ordinary abjection[22] of "Asian" women.

In doing this, Norico borrows a stylized version of "Western/global" vocabulary of the bargirl, strip, and drag routines and plays these acts in her bad girl style of excess in her girl culture-nation.[23] She performs "excess" mutations of seduction and play that cannot easily be contained and thus punished or sanctioned. Her mixture of Western-ish characters, ethnicked costumes, drag-like make-up, and musical and dance mixtures runs counter to nation-state culture. When Norico (in her Afro-do) does a karaoke singing routine of Vera Lynn's We'll Meet Again, (. . . don't know where, don't know when) while she wiggles her vinyl penis over the face of Bush, Jr. and then has Bush's photo face kiss Saddam Hussein's, the mix is layered with the ending of the 1964 film, *Dr. Stangelove* when the song was played with the bombs falling. Her girl culture takes on the "states" of nation and hegemony and queries the very foundations of power and the "national." Her commodified acts easily demonstrate the late-modern, even postcolonial nation as a compendium of "products" that are sold by the top "vendor of the nation."

Norico's flesh, a corporeal intersubjectivity, is multi-veiled, and cannot be touched in performance. Here is where "sex with nation" begins; her "ethnicked" costume removes her from "Japan," making her more available and more fun. In the next section her harem mask and top fall away with velcro-rip magic to the "real" breasts with Norico's signature not-bare act: she wears a loop of glittering chains between her breasts, each end attached (ouch) to each nipple. Out of her "veils" she is her signature OK Girl in her own disguise/persona (usually with a blonde wig, but here she keeps her Afro-do). She proceeds to dance more wildly shaking and snapping her "chains" off of her breasts, getting an audible wince from the spectators. The strip acts thus far are stock-acts, self-caressing, crotch shots, and bumps and grinds that show off Norico's body beneath the acts, but nevertheless in disguise. At this point, the action shifts to a more frenetic pace and music, Norico snaps herself into the bi-gender or bi-genital acts: putting on a pink vinyl penis (permanently at 45 or 60 degrees) and a breasts-and-abdomen set (with very extended nipples). She whips out her whip and abruptly turns from us as the classical music crescendos into techno-punk noise.

The Bad Girl Act: What To Do With The Weapons

In another act, Norico's attendants hold up a clothesline upstage, and with bumps and humps, Norico hangs up her "laundry." This section is an exquisitely complex choreography of nation-state succumbing to girl culture: vinyl sexual organs meeting photographic representations, mixed with gender, racial, and S/M corporeal mimicry. Norico first hangs up photographs, mug shots, of the global leaders in black and white. The War Boys: Saddam Hussein, Bush Jr., Tony Blair, Koizumi (the Japanese prime minister), North Korea's Kim Jong-II, and Israeli Prime Minister Ariel Sharon, are hung on this clothesline between the attendants, who are in drag outfits. The headshots are pinned on a rope, like laundry, or flags. Their faces wave at the bumps and grinds of Norico's flapping breasts and penis. First, she dances "for" them in boxing gloves doing street style and hip-hop and *eroporn* kickboxing. It is all staccato, erotic, and repetitive, like a loop of machine-gun hits and thrusts. She proceeds to dance with one or two of the photograph headshots, forcing her penis to their photo-faces with gleeful grins and oohs and aahs. Girl culture has sex with several nation heads at once! She then selects certain leaders, targeting Bush and Koizumi to butt with her swinging vinyl penis, just tossing the others into a pile to stomp on. Attendants put the headshots back on the line for round two.

In a sweeping mood change, Norico goes after the boys again. As I stated in the opening example, the music changes to the World War II classic "for the troops" song: *We'll meet again, don't know where, don't know when.* This time, Norico spits on several photograph faces and smears their faces together, a spit-kiss! Saddam licks Bush and Bush frenches Tony. Norico makes a girl-game of very high stakes. Turning to the Japanese Prime Minister, Koizumi, she crouches down back to audience and appears to "shit" on his face. When she stands again, she displays her abject act and presses Koizumi's shit-face into the other faces on the line. She presses them face to face, hangs them up side by side, their faces now marked with her excrement and spit. Dirty underwear? Girls' nation revenge? Twisted fun for the hegemonic heads of nations? A bad cosmetic surgery? Just the global girl dancer venting? I think it's more: Her finale takes on the ultimate cliché: a toy Uzi the high tech missile-gun of the "future-is-now" shock and awe war. Into the barrel, Norico thrusts a giant flower. The act ends in the ironic triumphant stance of modern warfare: gun over head, vinyl penis and breasts shivering and glistening, she blows us a kiss, singing the last lines:

> We'll meet again, don't know where, don't know when,
> But I know we'll meet again, some sunny day.
> Keep smiling through, just like you always do,
> 'Til the blue skies drive the dark clouds far away.

> So will you please say hello to the folks that I know,
> Tell them I won't be long.
> They'll be happy to know that as you saw me go,
> I was singing this song.[24]

Norico performs her own girl "culture nation": her own intersubjectivity that disrupts the "national" hegemonic order. "How to use the weapons" starts with the stripping of ethnicked racial layers, couples sex and gender with signs that slide, and then plays with bad girl deviant power.

Norico's performances of dresses, drag kings, and strip tease, which are all a part of global culture's oppression and commodity systems, create pleasure within their hybrid conditions of being bad girls in fantasy pop-Japan. These radical performed mutations take the globally marketed Japanese cute girl stuff and twist those products of consumption. Specifically in these song and dance acts examined here, the global pop club entertainment iconography is played sweetly, grotesquely, and ironically. The second twist is in the second example, in which the girl/woman "drags" global drag king/queen iconography into the Japanese girl scene and then merchandizes it for a United States tour. The "global" transfer of "drag" in Japan offers its own hybrid forms that Norico consumes and then performs in her deformed version of "drag" trans-girl-nation.

The circulation of female bodies is made strange through Norico's OK girl style in song and dance acts. Norico's OK (bad) Girl acts are part terror and pleasure. Norico's "What To Do With The Weapons" is a violent performance of Sex with the Global Nation: The bad girl shakes and aims her gun at the nation, "We'll meet again, don't know where, don't know when. . . ."[25] Norico's bad girl acts should cause tremors beyond the underground. Everything is not OK.

Notes

1. Norico Sunayama. English invitation for entering her installation, "Suffocating World," Art Tower Mito, 1999.
2. Amelia Jones, *Body Art: Performing the Subject* (Minneapolis: University of Minnesota Press, 1998), 152.
3. Peter Eckersall, "What Can't Be Seen Can Be Seen: Butoh Politics and (Body) Play," in *Body Shows*, ed. Peta Tait (Amsterdam and Atlanta, GA: Rodopi, 2000), 150.
4. Dennis Altman, *Global Sex* (Chicago: University of Chicago Press, 2001), 100.
5. Ibid.
6. Trinh T. Minh-ha, lecture, "Postcoloniality and Its Tools" seminar (September 2003), University of California, Berkeley.
7. Altman, *Global Sex*, 15.
8. Ibid., 50.
9. "Gura Gura Glamour Girl," *Montage* 007 (Summer 2000): 1–2.
10. "Norico Sunayama, OK Girls, Interview," *Montage* 007 (Summer 2000): 5.
11. "Norico Sunayama explains," *Montage* 007 (Summer 2000): 10
12. For a discussion of female performers and body/media in Dumb Type, see Katherine Mezur, "Fleeting Moments: The Vanishing Acts of Phantom Women in the Performances of Dumb Type," *Women and Performance: A Journal of Feminist Theory*, 12: 1 (#23 2001): 189–206.
13. Norico Sunayama, personal interview, Los Angeles, 2002. Norico insists that her caberet and club acts are entertainment and fun. There is no room here for a longer discussion on the Japanese concept of *asobi* (play) which is more complicated than the

Western fun-time idea. I suggest that "pleasure" may be closer to Norico's "entertainment" focus.

14. Larissa Hjorth, "Pop and Ma: The Landscape of Japanese Commodity Characters and Subjectivity," in *Mobile Cultures: New Media in Queer Asia*, ed. Chris Berry, Fran Martin, and Audrey Yue (Durham, NC: Duke University Press, 2003), 158–179.

15. Ibid., 174.

16. Jones, *Body Art*, 155.

17. Ibid.

18. Ibid., 153. Jones argues for the necessity of excess of strategies in speaking of Hannah Wilke's nude performance art and how these may be necessary for realizing "other" desires.

19. Norico prefers to be called by her first name, not her family name, Sunayama.

20. Ross Parker, and Hughie Charles, "We'll Meet Again," April 26, 2004, online at <http://ingeb.org/songs/wellmeet.html>; Vera Lynn, singer April 26, 2004, online at <http://www.lyricsxp.com/lyrics/w/we_ll_meet_again_vera_lynn.html>.

21. One example that clearly demonstrates this process is how most airlines use a young female "ethnicked" body to sell the nation's cultural hot spots and traditions: She may be the kimono girl (Japan), Golden crowned dancer girl (Indonesia), or brilliant sari-wrapped girl (India).

22. I am using some of the discourse on "abjection" that Karen Shimakawa has put forward in her book, *National Abjection*. While I am not looking at the same field of immigrant and nation, Norico uses methods of "abjection" and performs "abject" in a shifting "national/cultural identifying process." Karen Shimakawa, *National Abjection* (Durham, NC: Duke University Press, 2002), 3. Shimakawa also draws on Judith Butler's "abject" as theorized in *Bodies that Matter*, particularly her use of the "unlivable," (Judith Butler, *Bodies that Matter: On the Discursive Limits of Sex* [New York: Routledge, 1993], 3), which I find applicable to the nation/culture corporeality of Norico's performances or performed bodies.

23. Katherine Liepe-Levinson's book, *Strip Show: Performances of Gender and Desire, Gender in Performance* (London and New York: Routledge, 2002) and *Mobile Cultures: New Media in Queer Asia* ed. Chris Berry, Fran Martin, and Audrey Yue, both offer ways to analyze "acts" of desire, and the relationship of the market to those acts and their proliferation and audiences.

24. Ross Parker and Hughie Charles, "We'll Meet Again."

25. Ibid.

Afterword: AND some NOT SO BAD

Miriam Silverberg

We now know that notwithstanding the stellar example of Abe Sada *bad* can mean rightfully vengeful wronged women (the warp and weft of legends), unconventional make-up, and for the Japanese consumer in Paris—just bad shopping. In other words, not only can Japanese girls continue to change class just as they have under every imperial reign, some persist in what we in the bourgeois west have termed "unladylike" behavior, thereby breaking the gender line. Witness the cultured daughters of Tokugawa era entrepreneurs, or more recently, Empress Michiko rendered wraithlike (ghostly from years of monitoring by the Imperial Household Ministry, we must presume). Moreover, we have learned that not only sex can be altered; sex *and* race can change and must be added to the options available to the postmodern girl who would dare to be bad in late capitalist Japan.[1]

Now, at the turn of a new century, here in America, too many of us are still afraid to even use the word sex. Nervous newscasters are determined to misuse the term "gender." In America, when we hear that people have been "intimate," we know what they were really doing. And when we refer to geisha (with a smirk) and buy fictional memoirs of same, we think we presume to know what it's all about. We also know that bad girls relish sex (and we are both judgmental and jealous of this predilection). But we have also come to accept that often bad means good, that *girl* used to be a disrespectful term for woman but that now girl *power* means that girls need no longer be trapped in the culturally appropriate boxes designed for them. Now, girls (at least those fortunate to hear the admonition "you *go*, girl!") are on the move.

We are painfully aware that this was not always the case. Only the bad working-class girls got to wear the tight skirts and the dark, black eye make-up. The lucky ones got to meet their guys at the candy store and to revel, astride Harleys. And they were the ones, also, who were given entrée to the thrill of tragedy. Meanwhile, the good girls were back in their dorms, having their not sufficiently compensatory milk and cookies. (To talk about class, even if indirectly, is always, in some ways to talk about sex and the expression—or repression—of desire.)

We all know the good girl: good posture, good manners, good family. It is in this context that the good girl has a way of challenging our timidity regarding the danger of the universal. For while her prescribed and idealized comportment and deportment take on differing forms, a similarity cuts across times, places, and fashion statements. Take your whitest of *tabi*, footwear to be positioned pigeon-toed. Clearly Japan-specific, and at the same time but one of a legion of sartorial practices devised to highlight the female form by *strategically*, simultaneously covering surfaces of her body deemed threatening to male workaday piece of mind.

I am not interested in engaging with the good girl. Instead, I would like to remind my readers of the *not good* girls posing as good. We cannot refer to these figures as "bad" for fear of inviting the slightest of opprobrium; we do not want an interpretation of bad as good. Rather than defining "not good" let me illustrate with examples of some women close to home (and some too close to power). Condoleezza Rice—not good. And conversely, at the same time, in the same place, Barbara Boxer (for challenging Rice's veracity by using the new Secretary of State's own pro-war words)—bad.[2] Kathleen Harris—not good;[3] Lizzie Borden—not good;[4] the mother of the *Manchurian Candidate* (played by Angela Lansbury)—not good;[5] Cinderella's step-mother—not good.[6]

I will not pause here to elaborate on the class of girls who belong under the category of "not as bad as" (with bad in either its pejorative or affirmative meanings), although the pairing of the two literati ladies Murasaki Shikibu and Sei Shōnagon does come to mind. In this case the former is not as bad as the latter because the latter woman writer chose to ridicule members of their community.[7] But Murasaki Shikibu wasn't so bad. And this is the sort of bad girl I would have join the so many impressive bad girls whose histories are recorded in the pages of this book. We need to salute them as "not so bad!" We need to give grateful acknowledgment especially to the political bad girls who talked back, ignoring the directive given to girls through the ages. They were to be quiet, voiceless. And not only did these bold girls not keep their mouths shut, they also saw to it that their voices would be recorded, so that we could remember, in these opening years of the following century, their righteous (and rightful) anger at injustice. Not so bad.

Therefore, firstly, beware the bourgeois girl. These daughters of Meiji wielded the energy and the reserves that could only result from privilege. Rather than trying to determine the extent of their class-based privilege, it would be good to recognize that the well-bred schoolgirls with their hair ribbons and their uniform crisp *hakama* skirts were indeed the daughters of the class of capitalist men Marx referred to as revolutionary.[8] Not too bad. Therefore we need to take Hiratsuka Raichō at her word. If we do, we will hear her talking back to (of all people) Nora, the self-liberated heroine who slammed the door on her Scandinavian "doll's house" thereby abandoning children and a husband who used the names of small animals as terms of endearment. History tells us that Nora forged the way for Japanese and Chinese women to envision their own freedom. There was of course some concern about what would happen to our heroine after she had shut the door on her life. But Raichō went further. In an open letter politely addressed to Nora (one of the articles in an added section to the January 1912 issue of *Seitō*) its editor was relentless in venting her scorn for the heroine of *The Dollhouse*.

If it had really taken Nora eight years to realize that her husband was a total stranger to her then she was awfully slow on the uptake. Or she was a liar. In any case her constant begging was hypocritical. And, Hiratsuka went on to conclude, *Nora's husband was also a doll* (emphasis added). Not a bad line. Raicho saved her sharpest comments for last, in a suggestion that has most definitely not made its way into our histories. If Nora was not capable of leaving her double life in which all of her relationships were premised on lies; if she could not accept the miracle that a woman's life must be consistent and true, like a piece of religious music or like a poem, she must—and I here translate the blunt closing of the letter *verbatim*—"*go get a pistol or some poison. Good-bye.*" The letter was simply signed "H." No class solidarity there. Nonetheless it is true that the sexual politics set forth here do not leave the bourgeois drawing room. Not so bad.[9] It would seem that resentment articulated from a place of class privilege may in fact only go so far.

For a more systemic political analysis, we must witness a second bad girl talking back, this time in a courtroom wherein state officials attempt to reduce an instance of revolutionary will to an expression of interracial sexual servitude. Here I wish only to point out how Kaneko Fumiko kept charge of the testimony and thereby gave us her contemptuous analysis of the emperor system. Kaneko Fumiko, still a child at the time Hiratsuka was shutting the door in Nora's expectant face, was nobody's daughter, subject of no state, child of the colonies. She was captive to poverty and to her own relatives. Sent back to Japan, from Korea in 1919, as it happened, in the year of the Korean liberation uprising. Arrested four years later for conspiring together with her Korean common-law husband to assassinate the Emperor of Japan. Death by her own hands, by hanging, within three years.[10] We might develop the idea that Kaneko Fumiko was a Japanese incarnation of the sorceress whose story Hélène Cixous and Catherine Clément have traced in *The Newly Born Woman*.[11] As a woman, Kaneko's mythic yet historic existence was an anomaly outside of any acceptable narrative. She was marginalized from childhood because her family had not placed her name in the family registry, thereby denying her the necessary monitoring by the imperial state along with obligatory membership in the Japanese nation. She is most assuredly assured a place with the others who are abnormal. In Europe those would be the "madmen, deviants and neurotics; the women, drifters, jugglers, the tumblers" recalled by Cixous. It would take much more than these concluding afterthoughts to work out a place for a modern Japanese sorceress. When we do, let's not forget that the European sorceress suffers before an audience of men, nor that in Japan the mad were not altogether other.

But the Japanese Superior Court as similar male audience? Possibly, yet according to the spaces between the lines of Fumiko's recorded testimony, she appears not to have suffered, but to have relished the role of placing her accusations on exhibit. Such as her analysis of the three-tiered national social formation: beneath the imperial family (for which she felt pity), the ministers of state. Below them, the undifferentiated people. Suffice it to say that the accused on more than one occasion expressed irritation with the repetitious nature of the queries as she defined her relationship with Pak. The two were political comrades, was her emphasis. No, her political involvement was not caused by sexual excitement. (I will not here stoop to the level of elaborating on the notorious photograph of Kaneko Fumiko sprawled on Pak Yeol's lap, leaning back, positioned between his legs. Suffice it to say that he most definitely appears to be posing for the press as he looks right at camera. While she appears comfortable, she is too busy to look. Feet planted on the ground, she is reading.) Fumiko ended her first day of testimony on October 25, 1923 by explaining that one goal of her organization was to provide a base for free, "direct actions," and "we wanted to rile up honored officials like you people."[12]

A designation of Kaneko Fumiko as sorceress might in fact be appropriate, since she places herself entirely outside of the state-society: "I am not a person of Japan nor am I a person belonging to any other country." It is not that she does not want to be a member of a family-state that will not let her into the family. She has not been admitted and she most definitely does not wish it so. This is all written down for us by her in her poetry, her memoir, and her words in more than twenty separate occasions of court testimony. In closing, I merely want to recount some of her language, in order to try to convey, through the mediations of translation and of interpretation, the presence of this young girl.

Instance One: When she was asked the name of the magazine put out by *Futeisha*, the organization of Nihilists co-founded by Fumiko and Pak. They had not been allowed to use the term *Futeisenjin* (unruly Korean). This was not surprising, given the more than six thousand Koreans who had been massacred after the recent earthquake. To refer to out-of-line Koreans as *futei* (unruly) thereby appropriating the rallying cry of the neighborhood vigilante groups who had waylaid the thousands of innocent immigrants would have been hardly appropriate. They had therefore changed the title to *Futoisenjin*—meaning thick or fat Korean. Only a bad girl with flair could have effected that change!

Instance Two: When she answered the query: How did she refer to the Emperor?

"... the Invalid"
  And the Crown Prince?—
  "Rich Boy"
  And police officials?
  "Those guys at the police department are guard dogs for the '*bourgeois*' so we called them bulldogs (pun on bourgeois dogs) or little doggies."

The accused illustrated the use of her lexicon for the court by volunteering the following, at one of the numerous junctures when she was asked about plans made with Pak Yeol: "One time Pak said he'd once been a mailman so why couldn't he, at a time when the rich boy's procession was passing by, put on a uniform in order to deceive the bulldogs, and then throw a bomb at Rich Boy."

As I have already stated, such bad girls are aware that they are speaking for the record. Thus we must presume that the thirteenth testimony, on May 21, 1924 which took place at the request of the accused, was in part for our benefit: Fumiko explained that she had taken to heart the admonition received after she had used the term "critter" in reference to "getting a rich boy." She wished to change her testimony so as not to express special reverence but she also did not intend abuse. The record should read *not* that she had approved of getting one of those rich boy critters, but that she had said "It would be good to kill one Crown Prince." Not a bad statement for a bad girl. Really not so bad.

Moving on: Returning to the Repressed

I have come to the end of my Afterword but these concluding comments suggest a new discussion. This would entail relating all the bad girls in this book to each other. As an historian I cannot give up the diachronic. As an historian who respects the power of myth and of illusion I cannot give up a predilection for suggesting what cannot be proven. It is in this light that I suggest that Japanese bad girls be linked in a karmic series of dance partners, linking arms and moving forward to the next. (As opposed to the separated, organized motion of the *bon odori* with its unifying drum beat—the karmic dance would be much more unruly). They're always trying to put bad girls down. Let's keep them going by naming each bad girl a "returnee" reincarnated from another bad girl. Of course, it's hard to stop bad girls in motion, they will refuse to stay in place in Japan. The connected possibilities are beyond imagining.

You go girls. And you keep going.

Notes

1. At the risk of being accused of poor scholarship I have herein chosen to keep citations at a most general level. Readers who are desirous of more detail may want to think of this Afterword as a form (for lack of a more established term) of historical journalism. I do request that they treat it as what I have elsewhere termed *associative* history. Associated history highlights connections between cultures as opposed to comparing them, as in the case of historical writing premised on modernization theory.
2. See *New York Times* archive: "Transcript of remarks between Boxer and Rice," from Congressional confirmation hearings. Refer also to "Take Action hold Condoleeza Rice Accountable," *New York Times* archive and *New York Times* January 20, 2005, Thursday. Carl Hulse, "Boxer Is Loudest Voice of Opposition to Rice Nomination," *New York Times*, Section A, Page 6, Column 3.
3. Refer to the election won by Al Gore.
4. Even allowing for mental distress, murders of family members with an axe as the weapon of choice was not good.
5. The original movie, not the remake.
6. With the caveat that all step-mothers, by definition are not good.
7. See Biographies, Female Heroes of Asia: Japan, Murasaki Shikibu, <http://www.womeninworldhistory.com/heroine9.html>. See also Sei Shōnagon Biography at <http:// www.infoplease.com/ce6/people/A0844331.html>. The fact that the former and not the latter gained entry into the "Female Heroes of Asia" site contributes to proving my point that Sei Shōnagon was worse. For additional hagiography see "Samurai Sisters" in the same series.
8. Karl Marx, *The Communist Manifesto.* <http://www.marxists.org/archive/marx/works/1848/communist-manifesto/index.htm>.
9. Hiratsuka Raichō "Nora-san ni," *Seitō* (January 1912), 133–141.
10. For an intriguing, in-depth study see Helene Bowen Raddeker, *Treacherous Women of Imperial Japan: Patriarchal Fictions, Patriarchal Fantasies* (New York: Routledge, 1997). Kaneko Fumiko's prison memoir has been translated as *The Prison Memoirs of a Japanese Woman.* See the introduction by Mikiso Hane (Armonk, NY: M.E. Sharpe, 1991).

11. Hélène Cixous and Catherine Clément, "The Newly Born Woman," trans. Betsy Wing, *Theory and History of Literature*, Vol. 24 (Minneapolis: University of Minnesota Press, 1986).
12. For a complete record of testimony given by Kaneko Fumiko and by Pak Yeol, see *Gendaishi shiryō*, Vol. 3 (Tokyo: Misuzu Shobō, 1988).

Bibliography

Art Exhibits, Film, Music, Performances, and Television Sources

Ai no Koriida (In the realm of the senses). Film directed by Ōshima Nagisa, 1976.

"Burando (Brand)." Serialized television drama produced and directed by Wada Kō. Japan, 2000.

Frank Chickens. "We Are Ninja (Not Geisha)." LP Record Album. Produced by Kaz Records, 1984.

Heaven 17. "Geisha Boys and Temple Girls." Album title: *Pavement Side*. Produced by Virgin Records, 1981.

Ippai no kakesoba (A bowl of buckwheat noodles). Film directed by Nishikawa Katsumi, 1992.

Jitsuroku Abe Sada (The true story of Abe Sada). Film directed by Tanaka Noboru, 1975.

Lynn, Vera. *Lynn, Vera Singer*, 2004 [cited 26 April]. Available from <http://www.lyricsxp.com/lyrics/w/we_ll_meet_again_vera_lynn.html>

Nagasaki University Database of Old Photographs of the Bakumatsu-Meiji Period. Online at <http://oldphoto.lb.nagasaki-u.ac.jp/unive/>

"New Cosmos of Photography" (Shashin shin-seiki) sponsored by Canon, 1995 Gallery Exhibition Report. Online at <http://www.canon.com/scsa/newcosmos/gallery/1995/>

Parker, Ross and Hughie Charles. *We'll Meet Again*, 2004 [cited 26 April]. Available from <http://ingeb.org/songs/wellmeet.html>

Revolutionary Girl Utena (Shōjo kakumei Utena). Animation. Japanese TV series April 2–December 24, 1997; movie, 1999.

Sunayama, Norico. "English invitation for entering her installation, 'Suffocating World.' " Art Tower Mito, 1999.

Women in Japan: Memories of the Past, Dreams for the Future. Documentary film produced and directed by Joanne Hershfield and Jan Bardsley, 2002.

Manga, Magazine, Newspaper, and Websites (no author given)

"Abe Sada-Sakaguchi Ango taidan (Dialogue: Abe Sada-Sakaguchi Ango)." *Zadan* 1, no. 1 (1947): 30–35.

Biographies, Female Heroes of Asia: Japan, Murasaki Shikibu. Online at <http://www.womeninworldhistory.com/heroine9.html>

"10 dai yamamba gyaru osoru beki bi-ishiki dai chōsa!! (Big survey of aesthetic taste: Teenage witch girls should be worried!)." *Spa!*, September 1, 1999, 136–139.

"Dansei 100 nin no shōgenshū kawaii onna v. kawaikunai onna. (One hundred collected male testimonies on women who are cute, women who are not cute)." *Say*, No. 229, July 2002, 115–118.

"E-Girl no akogare mono (Things egg girls desire)." *Egg*, Vol. 85, November 2003, 87.

"E-Girl no puri chō Show (E-Girl's print club album show)." *Egg*, Vol. 85, November 2003, 74–77.

"E-Jump photo mail," *Egg*, Vol. 92, June 2004, 86–87.

"Ego shame (Ego cell-phone photos)." *Ego System*, Vol. 49, July 2004, 75–77.

"Fotojenikku teku oshiechaimasu (Teaching you photogenic techniques)." *Cawaii*, May 2003, 147–149.

"Ganguro musume no sugao ga mitai (We want to see the real faces of our black face daughters!)." *Shūkan hōseki*, April 14, 2000, 54.

Gendai nikan. Online at <http://www.bookreview.ne.jp/list.asp>

"Girl Gangs Brawl with Molotov Cocktails." *Mainichi Shimbun*, 26 June, 2001. Online at <http://mdn.mainichi.co.jp/news/archive/200106/26/20010626p2a00m0fp006002c.html>

"Gura gura glamour girl." *Montage 007*, Summer 2000, 1–15.

"Gyaru + animaru = gyanimaru zōshoku (Gyaru+animal = Gyanimal breeding)." *Focus*, January 1998, 10–11.

"Heisei jungle tanken—Kashima kyōju, 'nichijōseikatsu no hikyō' o motomete kyō mo iku 3 (Professor Kashima explores the Heisei jungle in search of 'uncharted regions of everyday life' 3)." *Gendai*, February 2002, 326.

"Hermès Opens Glitzy Shop in Ginza." *Japan Today*, June 28, 2001, online at <japantoday.com>

"Ima koso yamamba gyaru mukei bunka sai ni shite (Witch girls must be classified as national cultural property before it is too late)." *Shūkan Playboy*, May 2, 1999, 198–201.

"Japan's Luxury Product Sales Increase." *IGN Global Marketing Newsletter*, September 1999. Online at <http://www.pangaea.net/IGN/news0051.htm>

"Japan's Derided Women Photographers are Earning New Recognition," *AsiaWeek*, August 13, 1999. Online at <http://www.asiaweek.com/asiaweek/99/0813/feat2.html>

"Jogakusei daraku monogatari (Tales of degenerate schoolgirls)." *Yokohama Mainichi*, September 20–November 17, 1905.

"Jogakuseikan (Views of the schoolgirl)." *Shumi* (Tastes), March 1908.

"Kogyaru kara yamamba e: Atsuzoko, ganguro, bakkaka: Kogyaru wa shidai ni kuroku, kitanaku (From *kogyaru* to witches, platform boots, black face, idiot-ization: Kogyaru on the darker and dirtier program)." *Spa!*, July 2003, 26–27.

Komikku Amour, vol. 14, no. 8 (164), August 2003.

"Kyabukura wa yamamba mitai busu no ni natta (Cabaret clubs have become lairs for those ugly witches)." *Shūkan Post*, October 8, 1999, 63.

New York Times Archive. Online at <http://pqasb.pqarchiver.com/nytimes/advanced search. html>

"Norico Sunayama, OK Girls, Interview." *Montage 007*, Summer 2000, 5.

"Norico Sunayama explains." *Montage 007*, Summer 2000, 10.

"Ondanka no eikyōka? Soshite ichiji no boom ka? Shinka ka? Nippon wakamono no Latin-ka genzō o saguru! (Is it the influence of global warming, evolution, or a passing trend? Probing the 'Latinization' of Japanese youth!)." *Spa!*, February 9, 2000, 47–53.

"Otto ni naisho no shakkin de jiko hasan wa kannō? (Is it possible to go bankrupt because of loans taken without telling your husband?)." *Onna no toraburu*, October 1, 2003, 165–184.

"Pages Torn from the Memoirs of a Geisha. Mineko Distances Herself from Sayuri." *The Japan Times*, March 12, 1999.

"Purikura guranpuri (Print club grand prix)." *Ego System*, Vol. 49, July 2004, 84.

"Purikura kishu chō tettei hikaku (Super complete comparison of different print club equipment)." *Popteen*, June 2004, 196–201.

"Saikin purikura pōzu mihon (Sample album of recent print club poses)." *Cawaii*, November 2003, 141–145.

Sei Shōnagon Biography. Online at <http://www.infoplease.com/ce6/people/A0844331. html>

"Shōjotachi no shingo, ango, ryūkogo (New girls' words, codes, and slang)." *Dacapo*, October 15, 1997.

"Tadaima AV ni mo zōshokuchū ganguro, yamamba tte ii? (Are we going to have even more of these witch and black face porno videos!?)." *Focus,* March 8, 2000, 24–25.

"Toragyaru osorubeki enjo kōsai: Joshikōsei saisentan rupo (Terrifying Drunken Tiger Girls compensated dating: A report from the high school girl frontline)." *AERA,* 9:16 (April 15, 1996), 62.

"Uchira no gakkō no abunai sensei (The unreliable teachers at our school)." *Egg*, Vol. 92, June 2004, 54.

"Yabapuri taishō happyō (Presenting first place repulsive print clubs)." *Cawaii*, January 2003, 147–149.

Books and Articles

Abu-Lughod, Lila. "The Romance of Resistance." *American Ethnologist* 17, no. 1 (1990): 41–55.

Adachi, Jiro. "How Q Found Her Groove." *The New York Times*, January 30, 2005.

Aguiar, Sarah Appleton. *The Bitch Is Back: Wicked Women in Literature.* Carbondale and Edwardsville: Southern Illinois University Press, 2001.

Aida Makoto. *Mutant Hanako.* Tokyo: ABC Shuppan, 1999.

Aketa Tetsuo. *Nihon hanamachi shi* (History of Japan's *hanamachi*). Tokyo: Yūzankaku Shuppan, 1990.

Altman, Dennis. *Global Sex.* Chicago: University of Chicago Press, 2001.

Angst, Linda. "The Sacrifice of a Schoolgirl: The 1995 Rape Case, Discourses of Power and Women's Lives in Okinawa." *Critical Asian Studies* 33, no. 2 (2001): 243–264.

Aoki, Michiko Y. "Empress Jingū: The Shamaness Ruler." In *Heroic with Grace: Legendary Women of Japan*, edited by Chieko Mulhern, 3–39. New York: M. E. Sharpe, Inc., 1991.

Ariyoshi Sawako. *Hishoku* (Colorless). Tokyo: Kōdansha, 1967.

Atkins, E. Taylor. *Blue Nippon: Authenticating Jazz in Japan.* Durham, NC: Duke University Press, 2001.

Awazu Kiyoshi et al., eds. *Abe Sada: Shōwa jū-ichi nen no onna* (Abe Sada: Women of the year in Shōwa 11). Tokyo: Tabatake Shoten, 1976.

Bardsley, Jan. *The Bluestockings of Japan: New Women Essays and Fiction from Seitō, 1911–1916.* Ann Arbor, MI: Center for Japanese Studies, 2006.

Beichman, Janine. *Embracing the Firebird: Yosano Akiko and the Rebirth of the Female Voice in Modern Japanese Poetry.* Honolulu: University of Hawai'i Press, 2002.

Bernstein, Gail Lee, ed. *Recreating Japanese Women, 1600–1945.* Berkeley: University of California Press, 1991.

Berry, Chris, Fran Martin, and Audrey Yue, eds. *Mobile Cultures: New Media in Queer Asia.* Durham, NC: Duke University Press, 2003.

Blacker, Carmen. *The Catalpa Bow: A Study of Shamanistic Practices in Japan.* London: Allen & Unwin, 1975.

Borgen, Robert and Marian Ury. "Readable Japanese Mythology: Selections from *Nihon Shoki* and *Kojiki*." *Journal of the Association of Teachers of Japanese* 24, no. 1 (1991): 61–97.

Briles, Judith. "Banish Geisha Nursing (and Other Female-Dominated Professions)." In *The Briles Report on Women in Healthcare: Changing Conflict to Collaboration in a Toxic Workplace*, 201. San Francisco: Jossey-Bass, 1994.

Burns, Lori and Melisse Lafrance. *Disruptive Divas: Feminism, Identity and Popular Music.* New York: Routledge, 2002.

Butler, Judith. *Bodies That Matter: On the Discursive Limits of "Sex."* New York: Routledge, 1993.

———. "The Force of Fantasy: Feminism, Mapplethorpe, and Discursive Excess." In *Feminism and Pornography*, edited by Drucilla Cornell, 487–508. Oxford and New York: Oxford University Press, 2000.

Canell, Fenella. *Power and Intimacy in the Christian Philippines.* Cambridge: Cambridge University Press, 1999.

Cixous, Hélène and Catherine Clément. Translated by Betsy Wing. *Theory and History of Literature*, Vol. 24. Minneapolis: University of Minnesota Press, 1986.

Chalfen, Richard and Mai Marui. "Print Club Photography in Japan: Framing Social Relationships." *Visual Sociology* 16, no. 1 (2001): 55–77.

Chandler, Clay and Cindy Kano. "Recession Chic." *Fortune*, September 29, 2005, 52.

Cherry, Kittredge. *Womansword: What Japanese Words Say About Women.* Tokyo: Kodansha International, 1987.

Choy, Catherine Ceniza. *Empire of Care: Nursing and Migration in Filipino American History.* Durham, NC: Duke University Press, 2003.

Coaldrake, A. Kimi. "Female *Tayū* in the *Gidayū* Narrative Tradition of Japan." In *Women and Music in Cross-Cultural Perspective*, edited by Ellen Koskoff, 151–161. Urbana and Chicago: University of Illinois Press, 1989.

Cobb, Jodi. *Geisha.* New York: Alfred A. Knopf, 1995.

Cohen, Phil. "Subcultural Conflict and Working-Class Community." In *Working Papers in Cultural Studies* 2, 5–53. Birmingham: CCCS, 1972.

Commission on Filipinos Overseas. "Numbers of Registered Filipino Fianc(e)es/Spouses of Foreign Nationals by Major Country of Destination: 1989–2001." Online at <http//www.cfo.gov.ph>

Connell, Ryann. "Bold Teen Babes Flash Full Body Flesh for the Porn Print Pic," Mainichi Daily News Interactive, August 27, 2003. <http:mdn.mainichi.co.jp/waiwai/0308/0827.eropuri.html>

Copeland, Rebecca L. *Lost Leaves: Women Writers of Meiji Japan.* Honolulu: University of Hawai'i Press, 2000.

Cornyetz, Nina. "Fetishized Blackness: Hip Hop and Racial Desire in Contemporary Japan." *Social Text* 41 (1994): 113–139.

———. "Power and Gender in the Narratives of Yamada Eimi." In *The Woman's Hand: Gender and Theory in Japanese Women's Writing*, edited by Paul Gordon Schalow and Janet Walker, 425–457. Stanford: Stanford University Press, 1996.

Cowie, Elizabeth. "Pornography and Fantasy: Psychoanalytic Perspectives." In *Sex Exposed: Sexuality and the Pornography Debate*, edited by Lynne Segal and Mary McIntosh, 132–152. London: Virago Press, 1992.

Dalby, Liza Crihfield. "Courtesan and Geisha: The Real Women of the Pleasure Quarter." In *The Women of the Pleasure Quarter*, edited by Elizabeth de Sabato Swinton, 67–86. New York: Hudson Hills Press, 1996.

de Mente, Boyd. *Some Prefer Geisha: The Lively Art of Mistress-Keeping in Japan.* Rutland, VT: Charles E. Tuttle, Inc., 1966.

Dollase, Hiromi Tsuchiya. "Early Twentieth Century Japanese Girls' Magazine Storics: Examining *Shōjo* Voice in *Hanamonogatari* (Flower tales)." *Journal of Popular Culture* 36, no. 4 (2003): 724–755.

Dōmei Tsūshinsha. *Nihon no bukka to fūzoku 130 nen no utsurikawari: Meiji gannen~Heisei 7 nen* (130 years of Japan's changing commodity prices and customs from 1868–1995). Tokyo: Dōmei Tsūshinsha, 1997.

Douglass, Mike. "The Singularities of International Migration of Women to Japan." In *Japan and Global Migration: Foreign Workers and the Advent of a Multicultural Society*, edited by Mike Douglass and Glenda S. Roberts, 99–120. London: Routledge, 2000.

Dower, John W. *War without Mercy: Race and Power in the Pacific War*. New York: Pantheon Books, 1986.

———. *Embracing Defeat: Japan in the Wake of World War II*. New York: W.W. Norton & Co./New Press, 1999.

Eckersall, Peter. "What Can't Be Seen Can Be Seen: Butoh Politics and (Body) Play." In *Body Shows*, edited by Peta Tait, 145–151. Amsterdam and Atlanta, GA: Rodopi, 2000.

Egusa Mitsuko. *Watashi no shintai, watashi no kotoba: Jendā de yomu Nihon kindai bungaku* (My body, my words: Reading gender in modern Japanese literature). Tokyo: Kanrin Shobō, 2004.

Enchi Fumiko. *Masks*. Translated by Juliet Winters Carpenter. New York: Random House, 1983.

———. *A Tale of False Fortunes*. Translated by Roger K. Thomas. Honolulu: University of Hawai'i Press, 2000.

Erino Miya. "Redikomi no orugasumu ga josei no sei ishiki o kaeta? (Has the ladies' comic's orgasm changed women's sexual consciousness?)." *Bessatsu Takarajima* 30 (August 1994): 130–133.

Ferrell, Robin. "The Pleasures of the Slave." In *Between Psyche and Social: Psychoanalytic Social Theory*, edited by Kelly Oliver and Steve Edwin, 19–27. Lanham, MD: Rowman and Littlefield Publishers, Inc, 2002.

Field, Norma. "Somehow: The Postmodern as Atmosphere." In *Postmodernism and Japan*, edited by Masao Miyoshi and H.D. Harootunian, 169–188. Durham, NC: Duke University Press, 1989.

Fiske, John. "Cultural Studies and the Culture of Everyday Life." In *Cultural Studies*, edited by Cary Nelson, Lawrence Grossberg, and Paula Treichler, 154–173. New York: Routledge, 1992.

Flax, Jane. "Women Do Theory." *Quest* 5, no. 1 (1979): 20–26.

Frederick, Sarah. "Sisters and Lovers: Women Magazine Readers and Sexuality in Yoshiya Nobuko's Romance Fiction." *AJLS Proceedings* 5 (Summer 1999): 311–320.

———. "Bringing the Colonies 'Home': Yoshiya Nobuko's Popular Fiction and Imperial Japan." In *Across Time & Genre: Reading and Writing Japanese Women's Texts*. Conference Proceedings, edited by Janice Brown and Sonja Arntzen, 61–64. Edmonton: The University of Alberta, 2002.

Frühstück, Sabine. *Colonizing Sex: Sexology and Social Control in Modern Japan*. Berkeley: University of California Press, 2003.

Fudge, Rachel. "Grrrl: You'll Be a Lady Soon." *Bitch: Feminist Response to Pop Culture* 14 (2001): 22–24.

Fujimoto Yukari. "Onna no yokubō no katachi: Rediisu komikku ni miru onna no seigensō (The shape of female desire: Female sexuality as seen in ladies' comics)." In *Nyū feminizumu rebyū 3: Porunogurafi:Yureru shisen no seijigaku* (New Feminism Review 3: Pornography: The politics of a wavering gaze), edited by Shirafuji Kayako, 70–90. Tokyo: Gakuyō Shobō, 1992.

Fukushima Akira, Nakata Osamu, and Kogi Sadataka. *Nihon no seishin kantei* (Psychological examinations in Japan). Tokyo: Misuzu Shobō, 1973.

Gaonkar, Dilip. "On Alternative Modernities." *Public Culture* 11, no. 1 (1999): 1–18.

Garon, Sheldon M. *Molding Japanese Minds: The State in Everyday Life*. Princeton, NJ: Princeton University Press, 1997.

Golden, Arthur. *Memoirs of a Geisha*. New York: Alfred A. Knopf, 1997.

Guerrilla Girls. *Bitches, Bimbos, and Ballbreakers: The Guerrilla Girls' Illustrated Guide to Female Stereotypes*. New York: Penguin Books, 2003.

Hagi Kazuaki. "Yokatta, yokatta . . . dake dewa (Fine, fine . . . alone isn't enough)." *Asahi Shimbun*, October 25, 1987, 20–21.

Harootunian, H. D. *History's Disquiet: Modernity, Cultural Practice, and the Question of Everyday Life*. New York: Columbia University Press, 2000.

Hayashi Mariko. *Kosumechikku* (Cosmetics). Tokyo: Shōgakkan, 2002.

Hayashi Sawako. "Toshokan shiryō to shite no taishū jidōbungaku o kangaeru: Yoshiya Nobuko no shōjo shōsetsu o rei ni (Considering popular children's literature from the perspective of library resources: The case of Yoshiya Nobuko)." *Ōtani Joshidaigaku kiyō* 35, no. 1.2 (2001): 186–206.

Hebdige, Dick. *Subculture, The Meaning of Style*. London: Methuen, 1979.

Henshall, Kenneth G. *The Quilt and Other Stories by Tayama Katai*. Tokyo: University of Tokyo Press, 1981.

Herbert, Wolfgang. *Foreign Workers and Law Enforcement in Japan*. London: Kegan Paul International, 1996.

Hiratsuka Raichō. " Nora-san ni (To Ms. Nora)." *Seitō* 2:1 (January 1912): 133–141.

Hirota, Aki. "Image-Makers and Victims: The Croissant Syndrome and Yellow Cabs." *U.S.-Japan Women's Journal English Supplement* 19 (2000): 83–121.

Hjorth, Larissa. "Pop and Ma: The Landscape of Japanese Commodity Characters and Subjectivity." In *Mobile Cultures: New Media in Queer Asia*, edited by Fran Martin, Chris Berry, and Audrey Yue, 158–179. Durham: Duke University Press, 2003.

Honda Masuko, ed. *Shōjo-ron* (Girl theory). Tokyo: Seikyūsha, 1988.

———. *Jogakusei no keifu: Saishoku sareru Meiji* (A chronology of the schoolgirl: The Meiji period in color). Tokyo: Seidosha, 1990.

———. *Ibunka to shite no kodomo* (The alien culture of children). Tokyo: Chikuma Gakugei Bunko, 1992.

Hon no Mori Henshū-bu, ed. *Abe Sada Jiken chōsho zenbun* (A complete record of the Abe Sada incident). Tokyo: Kosumikku Intōnashonaru, 1997.

Horiba Kiyoko. *Seitō no jidai: Hiratsuka Raichō to atarashii onna-tachi* (The era of the Bluestockings: Hiratsuka Raichō and the new women). Tokyo: Iwanami Shoten, 1988.

Horinouchi Masakazu. *Abe Sada shōden* (The true biography of Abe Sada). Tokyo: Jōhō SentāShuppan Kyoku, 1998.

Horney, Karen. *Feminine Psychology*. New York: Norton, 1967.

Hoshino Sumire. *Gendai jogakusei hōkan* (Modern schoolgirl thesaurus) Vol. 3 of *Kindai Nihon seinenki kyōkasho sōsho* (Library of modern Japanese adolescent schoolbooks), 16 vols. Tokyo: Nihon Tosho Sentā, 1992.

Hulse, Carl. "Boxer Is Loudest Voice of Opposition to Rice Nomination," *New York Times*, January 20, 2005, Section A, Page 6, Column 3.

Ivy, Marilyn. *Discourses of the Vanishing: Modernity, Phantasm, Japan*. Chicago: University of Chicago Press, 1995.

Iwabuchi Junko, ed. *"Danna" to asobi to Nihon bunka* (Patrons, recreation and Japanese culture). Tokyo: PHP Kenkyūjo, 1996.

Iwasaki Mineko with Rande Brown. *Geisha, A Life*. New York: Atria Books, 2002.

Izbicki, Joanne. "The Shape of Freedom: The Female Body in Post-Surrender Japanese Cinema." *U.S.-Japan Women's Journal English Supplement* 12 (1997): 109–153.

Johnston, William. *Geisha, Harlot, Strangler, Star: A Woman, Sex, and Morality in Modern Japan*. New York: Columbia University Press, 2004.

Jolivet, Muriel. *Japan: The Childless Society?* Translated by Anne-Marie Glasheen. London: Routledge, 1997.

Jones, Amelia. *Body Art: Performing the Subject*. Minneapolis: University of Minnesota Press, 1998.

Jones, Gretchen. "'Ladies' Comics': Japan's Not-so-Underground Market in Pornography for Women." *U.S.-Japan Women's Journal English Supplement* 22 (2002): 3–31.

Joss, Frederick. *Of Geisha and Gangsters: Notes, Sketches, and Snapshots from the Far East*. London: Odham's Press, 1962.

Jowan, Dinah (romanized from Japanese). "Interview: Watashi no Japayuki taiken (Interview: My experience as a *Japayuki*)." *Bessatsu Takarajima* 54 (1986): 113–121.

Juffer, Jane. *At Home with Pornography: Women, Sex, and Everyday Life*. New York: New York University Press, 1998.

Kaibara, Ekken. *Women and Wisdom of Japan*. Translated by K. Hoshino. London: John Murray, 1905.

Kakuchi, Suvendrini. "Japan Economy: Despite Hard Times, Shopping Habits Die Hard." *Asia Times Online*, January 15, 1999.

Kaneko, Fumiko. *The Prison Memoirs of a Japanese Woman*. Translated by Mikiso Hane. Armonk, NY: M. E. Sharpe, 1991.

Kaneko Fumiko and Pak Yeol. *Gendaishi shiryō* Vol. 3: (modern history source material) Tokyo: Misuzu Shobō, 1998.

Karasawa Shun'ichi. "Redikomi no taiken tōkōsha wa, naze kaiinu to yatta hanashi made kokuhaku shitai no? (Why do readers of ladies' comics want to submit stories confessing they had sex with dogs?)." *Bessatsu Takarajima* 240 (1995): 192–202.

Kawahara Toshiaki. *Kōtaishi-hi Masako-sama* (Crown Princess Masako). Tokyo: Kōdansha, 1993.

Kawai Hayao. *The Japanese Psyche: Major Motifs in the Fairy Tales of Japan*. Translated by Hayao Kawai and Sachiko Reece. Woodstock, CT: Spring Publications, Inc., 1996.

Kawamoto Saburō. "Jiritsu suru 'onna Yoshikawa Eiji' (The independent 'female Yoshikawa Eiji')." *Shokun!*, January 2003: 276–277.

Kawamura Kunimitsu. *Otome no shintai: Onna no kindai to sekushuariti* (The maiden's body: Women's modernization and sexuality). Tokyo: Kinōkuniya Shoten, 1994.

———. *Sekushuariti no kindai* (The modernity of sexuality). Tokyo: Kōdansha, 1996.

Keene, Donald, ed. *Twenty Plays of the Noh Theater*. New York: Columbia University Press, 1970.

Kelsky, Karen. *Women on the Verge: Japanese Women, Western Dreams*. Durham, NC: Duke University Press, 2001.

Kingston, Maxine Hong. *The Woman Warrior: Memoirs of a Girlhood Among Ghosts*. New York: Random House, 1977.

Kinsella, Sharon, "Cuties in Japan." In *Women, Media and Consumption in Japan*, edited by Lise Skov and Brian Moeran, 220–254. Honolulu: University of Hawai'i Press; Richmond: Curzon, 1995.

———. "Fantasies of a Female Revolution in Male Cultural Imagination in Japan." In *Zap-Pa (Groupuscules) in Japanese Contemporary Social Movements*, edited by Kōso Sabu and Nagahara Yutaka. New York: Autonomedia, forthcoming 2005.

Kiriyama Hideki. "Kanojo ga 'kagekina manga' o yomu riyū (Why she reads 'hardcore manga')." *Purejidento* (1993): 192–195.

Klein, Susan. "Women as Serpent: The Demonic Feminine in the Noh Play *Dōjōji*." In *Religious Reflections on the Human Body*, edited by Jane Marie Law, 100–135. Bloomington: Indiana University Press, 1995.

Kohama Itsurō. "Shutai to shite no shōjo (Girl as subject)." In *Shōjo-ron*, edited by Honda Masuko, 85–106. Tokyo: Seikyūsha, 1988.

Kosugi Tengai. *Meiji Taishō bungaku zenshū dai 16 maki* (Vol. 16 in Meiji and Taisho literature collection). 60 vols. Tokyo: Shunyōdō, 1930.

Kosugi Tengai. *Makaze koikaze* (Demon winds love winds). 2 vols. Tokyo: Iwanami Shoten, 1999.

Koyama Shizuko. *Ryōsai kenbo to iu kihan* (The good-wife wise-mother paradigm). Tokyo: Keisō Shobō, 1991.

Kristof, Nicholas D. "In Japan, Brutal Comics for Women." *New York Times*, November 11, 1995, 6.

Laing, David. *One Chord Wonders: Power and Meaning in Punk Rock*. Philadelphia: Open University Press, 1985.

Larson, Joan Pross. *The Geisha Cookbook: Japanese Cookery for Americans*. Mount Vernon, NY: Peter Pauper Press, 1973.

Latham, Angela J. *Posing a Threat: Flappers, Chorus Girls, and Other Brazen Performers of the American 1920s*. Middletown, CT: Weslyan University Press, 2000.

LeBlanc, Robin. *Bicycle Citizens: The Political World of the Japanese Housewife*. Berkeley: University of California Press, 1999.

Lie, John. "The State as Pimp: Prostitution and the Patriarchal State in Japan in the 1940s." *Sociological Quarterly* 38, no. 2 (1997): 251–263.

Liepe-Levinson, Katherine. *Strip Show: Performances of Gender and Desire, Gender in Performance*. London and New York: Routledge, 2002.

Lippard, Lucy R. *From the Center: Essays on Women's Art*. New York: Dutton, 1976.

Mackie, Vera. *Feminism in Modern Japan: Citizenship, Embodiment and Sexuality*. Cambridge: Cambridge University Press, 2003.

Maesaka Toshiyuki. *Abe Sada shuki* (Memoirs of Abe Sada). Tokyo: Chūkōbunko, 1998.

Marran, Christine. "'Poison Woman' Takahashi Oden and the Spectacle of Female Deviance in Early Meiji." *U.S.-Japan Women's Journal English Supplement* 9 (1995): 93–110.

Maruta Kōji. "Giji-ibento to shite no enjo kōsai (Compensated dating as a pseudo-event)." *Osaka jogakuin tankidaigaku kiyō dai* 30 (2000): 209–222.

Marx, Karl. *The Communist Manifesto*. Online at <http://www.marxists.org/archive/marx/works/1848/communist-manifesto/index.htm>

Masaoka Geiyō. *Shimbunsha no rimen* (The other side of newspaper companies). Tokyo: Shinseisha, 1901.

Masubuchi Sōichi. *Kawaii shōkōgun* (Cuteness syndrome). Tokyo: NHK Shuppan, 1994.

Mathews, Gordon and Bruce White. *Japan's Changing Generations: Are Young People Creating a New Society?* London and New York: Routledge/Curzon, 2004.

Matlack, Carol with Diane Brady, Robert Berner, Rachel Tiplady and Hiroko Tashiro. "The Vuitton Money Machine." *Business Week International Editions*, March 22, 2004, 48.

McCormack, Gavan. *The Emptiness of Japanese Affluence*. Armonk, NY: M. E. Sharpe, 1996.

Mezur, Katherine. "Fleeting Moments: The Vanishing Acts of Phantom Women in the Performances of Dumb Type." *Women and Performance: A Journal of Feminist Theory* 12, no. 1 (#23 2001): 189–206.

Miller, Laura. "Crossing Ethnolinguistic Boundaries: A Preliminary Look at the *Gaijin Tarento* in Japan." In *Asian Popular Culture*, edited by John Lent, 162–173. New York: Westview Press, 1995.

———. "Bad Girls: Representations of Unsuitable, Unfit, and Unsatisfactory Women in Magazines." *U.S.-Japan Women's Journal English Supplement* 15 (1998): 31–51.

———. "Media Typifications and Hip *Bijin*." *U.S.-Japan Women's Journal English Supplement* 19 (2000): 176–205.

———. "Youth Fashion and Changing Beautification Practices." In *Japan's Changing Generations: Are Young People Creating a New Society?*, edited by Gordon Mathews and Bruce White, 83–97. London: Routledge/Curzon, 2004.

———. "Graffiti Photos: Expressive Art in Japanese Girls' Culture." *Harvard Asia Quarterly* 7, no. 3 (2004): 31–42.

———. "Those Naughty Teenage Girls: Japanese Kogals, Slang and Media Assessments." *Journal of Linguistic Anthropology* 14, no. 2 (2004): 225–247.

———. "You are Doing *Burikko*!: Censoring/Scrutinizing Artificers of Cute Femininity in Japanese." In *Japanese Language, Gender, and Ideology: Cultural Models and Real People*, edited by Janet Shibamoto Smith and Shigeko Okamoto, 146–165. Oxford: Oxford University Press, 2004.

Minamoto, Junko. "Buddhist Attitudes: A Woman's Perspective." In *Studies on the Impact of Religious Teachings on Women*, edited by Jeanne Becher, 154–171. Philadelphia, PA: Trinity Press International, 1991.

Ministry of Education. *Gakusei 80 nenshi* (Eighty-year history of the school system). Tokyo: Ōkurashō, 1954.

Ministry of Education, Science and Culture. *Japan's Modern Educational System: A History of the First Hundred Years*. Tokyo: Printing Bureau, Ministry of Finance, 1980.

Ministry of Health and Welfare. *Vital Statistics*. Tokyo: Ōkurashō, 1996, 2003.

Ministry of Justice. *Annual Report of Statistics on Legal Migrants*. Tokyo: Ōkurashō, 2003.

Mitamura Engyō. *Karyū fūzoku* (Culture of the geisha community), Vol. 27 of *Engyō Edo bunko* series, edited by Asakura Haruhiko. Tokyo: Chūō Kōronsha, 1998.

Miyadai Shinji. *Seifuku shōjotachi no sentaku* (The choice of the little girls in uniform). Tokyo: Kōdansha, 1994.

Miyake, Toshio. "Black Is Beautiful: Il Boum Delle Ganguro-Gyaru (The ganguro girl boom)." In *La Bambola E Il Robottone: Culture Pop Nel Giappone Contemporaneo* (Children and robots: Contemporary Japanese culture), edited by Allessandro Gomarasca, 111–144. Torino: Einaudi, 2001.

Miyashita Maki. *Heya to shitagi*. (Rooms and underwear). Tokyo: Shōgakkan, 2000.

Mizoguchi, Akiko. "Male-male Romance By and For Women in Japan: A History and the Subgenres of *Yaoi* Fictions." *U.S.-Japan Women's Journal English Supplement* 25 (2003): 49–75.

Modleski, Tania. *Loving with a Vengeance: Mass-Produced Fantasies for Women*. Hamden, CT: Archon Books, 1982.

Molony, Barbara. "Japan's 1986 Equal Employment Opportunity Law and the Changing Discourse on Gender." *Signs* (1995): 269–302.

Mori Nobuyuki. *Tokyo joshikō seifuku zukan* (The Tokyo high school girl uniform fieldbook). Tokyo: Kuritsusha, 1985.

Nagaike, Kazumi. "Perverse Sexualities, Perversive Desires: Representations of Female Fantasies and *Yaoi Manga* as Pornography Directed at Women." *U.S.-Japan Women's Journal English Supplement* 25 (2003): 76–103.

Nagashima Yurie. *Nagashima Yurie*. Tokyo: Fuga Shobō, 1995.

Nagata Mikihiko. "Abe Osada." *Fūfu seikatsu* (Conjugal life). September–August 1950–1951.

Nakamatsu Tomoko. "International Marriage through Introduction Agencies: Social and Legal Realities of 'Asian' Wives of Japanese Men." In *Wife or Worker?: Asian Women and Migration*, edited by Nicola Piper and Mina Roces, 181–201. Lanham, MD: Rowman & Littlefield, 2003.

Nakamura Usagi. *Gokudō-Kun manyūki* (Adventures of Gokudō-Kun). Tokyo: Kadokawa Bunko, 1991.

———. *Pari no toire s'il vous plait* (Parisian toitets, if you please). Tokyo: Kadokawa Bunko, 1999.

———. *Konna watashi de yokattara* (I'm yours if you like). Tokyo: Kadokawa Bunko, 2000.

Nakamura Usagi. *Shoppingu no joō* (The shopping queen). Paperback version. Tokyo: Bungei Bunko, 2001.

———. *Shoppingu no joō: Saigo no seisen!?* (The shopping queen: The last crusade!?). Tokyo: Bungei Shunjū, 2004.

———. *Datte, hosii n da mon!: Shakkin joō no binbō nikki* (But, I want it: The queen of loan's diary of impoverishment). Tokyo: Kadokawa Bunko, 1997.

Nakano Kōji. *Seihin no shisō* (Thoughts on pure and poor). Tokyo: Sōshisha, 1992.

Nakano, Midori. "Yamamba." *Japan Echo*, February 2000, 62–63.

Nakayama Akihiko "Shōsetsu 'Tokai' saiban no gingakei: Kūhaku no seijigaku (The galactic system of 'The City' trial: The politics of margins and voids)." In *Kindai shōsetsu no 'katari' to 'gensetsu'* ("Narrative" and "discourse" in the modern novel), edited by Mitani Kuniaki, 55–81. Tokyo: Yūseidō, 1996.

Narita, Ryūichi. "The Overflourishing Sexuality in 1920s Japan." In *Gender and Japanese History*, edited by Haruko Wakita, Anne Bouchy, Chizuko Ueno, and Gerry Yokota, 345–370. Osaka: Osaka University Press, 1999.

Noble, Marianne. *The Masochistic Pleasures of Sentimental Literature*. Princeton, NJ: Princeton University Press, 2000.

Noddings, Nel. *Women and Evil*. Berkeley: University of California Press, 1989.

Nolte, Sharon H. and Sally Ann Hastings. "The Meiji State's Policy toward Women, 1890–1910." In *Recreating Japanese Women, 1600–1945*, edited by Gail Lee Bernstein, 151–170. Berkeley: University of California Press, 1991.

Ogasawara, Yuko. *Office Ladies and Salaried Men: Power, Gender and Work in Japanese Companies* (Berkeley: University of California Press, 1998).

Ogura Chikako. *Kekkon no jōken* (Conditions for marriage). Tokyo: Asahi Shimbunsha, 2004.

Oguri Fūyō. *Seishun* (Youth). 3 vols. Tokyo: Iwanami Shoten, 1994.

Ohba Minako. "The Smile of the Mountain Witch." In *Stories by Contemporary Women Writers*, edited and translated by Noriko Mizuta Lippit and Kyoko Iriye Selden, 182–196. London and New York: M. E. Sharpe, 1982.

Ōnuma Shōji. *Minzoku* (Folk). Tokyo: Kawade Shobō Shinsha, 2001.

Orlotani, Benito. *The Japanese Theatre: From Shamanistic Ritual to Contemporary Pluralism*. Princeton: Princeton University Press, 1995.

Ōtsuka Eiji. *Shōjo minzokugaku: Seikimatsu no shinwa o tsumugu miko no matsuei* (The native ethnology of girls). Tokyo: Kōbunsha, 1989.

———. *Etō Jun to shōjo feminizumu-teki sengo subculture bungakuron* (Etō Jun and a girl-feminist critique of postwar culture). Tokyo: Chikuma Shobō, 1998.

Ōtsuki Kenji. *Abe Sada seishin bunsekiteki shindan* (The psychoanalytic diagnosis of Abe Sada). Tokyo: Tokyo Seishin Bunseki-gaku Kenkyūjo, 1936.

Otsuki Hisako. *Shinsen Tokyo joshi yūgaku annai* (New selection: A guide for girls studying in Tokyo). Vol. 9 of *Kindai Nihon seinenki kyōkasho shōsho* (Library of modern Japanese adolescent schoolbooks). 16 vols. Tokyo: Nihon Tosho Sentā, 1992. Reprint, 1905.

Philippine Women's League of Japan. *Pinays in Japan*, 2004. <Online at http://japan.co.jp/~ystakei/pwl1.html>

Pitchford, Nicola. "Reading Feminism's Pornography Conflict: Implications for Postmodern Reading Strategies." In *Sex Positives? The Cultural Politics of Dissident Sexualities*, edited by Thomas Foster, Carol Siegel, and Ellen E. Berry, 3–38. New York: New York University Press, 1997.

Portes, Alejandro and Ruben G. Rumbaut, eds. *Immigrant America: A Portrait*. Berkeley: University of California Press, 1996.

Raddeker, Helene Bowen. *Treacherous Women of Imperial Japan: Patriarchal Fictions, Patriarchal Fantasies*. New York: Routledge, 1997.

Radway, Janice. *Reading the Romance: Women, Patriarchy, and Popular Culture*. Chapel Hill, NC and London: University of North Carolina Press, 1984.

Ravitch, Diane. *The Language Police: How Pressure Groups Restrict What Students Learn*. New York: Alfred A. Knopf, 2003.

Robertson, Jennifer. *Takarazuka: Sexual Politics and Popular Culture in Modern Japan*. Berkeley and Los Angeles: University of California Press, 1998.

———. "Yoshiya Nobuko: Out and Outspoken in Practice and Prose." In *The Human Tradition in Modern Japan*, edited by Ann Walthall, 155–174. Wilmington, DE: Scholarly Resources, 2001.

———. "Blood Talks: Eugenic Modernity and the Creation of New Japanese." *History and Anthropology* 13, no. 3 (2002): 191–216.

Robertson, Pamela. *Guilty Pleasures: Feminist Camp from Mae West to Madonna*. Durham, NC: Duke University Press, 1996.

Roth, Katherine. "As 'Geisha Chic' Hits U.S., One, 87, Scoffs." *The Japan Times*, June 22, 1999.

Rowley, G.G. "Prostitutes Against the Prostitution Prevention Act of 1956." *U.S.-Japan Women's Journal English Supplement* 23 (2002): 39–56.

Russell, John. "The Black Other in Contemporary Japanese Mass Culture." In *Contemporary Japan and Popular Culture*, edited by John Treat, 17–40. Honolulu: University of Hawai'i Press; Richmond: Curzon, 1996.

Saeki Junko. *"Shiki" to "ai" no hikaku bunka shi* (A comparative cultural history of "lust" and "love"). Tokyo: Iwanami Shoten, 1998.

———. *Ren'ai no kigen* (The origins of love). Tokyo: Nihon Keizai Shinbunsha, 2000.

Sakaguchi Ango. "Abe Sada-san no inshō (My Impressions of Ms. Abe Sada)." *Zadan* 1, no. 1 (December 1947) 36–38.

Sakamoto Mimei. "Yappari onna wa baka datta (Women are fools after all)." *Shinchō 45* 16, no. 7 (1997): 234–243.

Sakuma Rika. "Shashin to josei (Woman and photography)." In *Onna to otoko no jikū: Nihon no joseishi saikō* (Space-time of men and women. Redefining Japanese women's history), edited by Kōno Nobuko et al., Vol. 2: 187–237. 5 vols. Tokyo: Fujiwara Shoten, 1995–1998.

Sapir, David J. "On fixing Ethnographic Shadows." *American Ethnologist* 21, no. 4 (1994): 867–884.

Sato, Barbara Hamill. *The New Japanese Woman: Modernity, Media, and Women in Interwar Japan,* Durham, NC: Duke University Press, 2003.

Scharf, Frederic, Sebastian Dobson, and Anne Nishimura Morse. *Art and Artifice: Japanese Photographs of the Meiji Era*. Boston: Museum of Fine Arts, 2004.

Seaman, Amanda. *Bodies of Evidence: Women, Society, and Detective Fiction in 1990s Japan*. Honolulu: University of Hawai'i Press, 2004.

Sei Shōnagon. *The Pillow Book of Sei Shōnagon*. Translated and edited by Ivan Morris. New York: Columbia University Press, 1991.

Sekine Hiroshi. *Sekine Hiroshi shi-shū: Abe Sada* (Collection of poems by Sekine Hiroshi: Abe Sada). Tokyo: Doyōbijutsusha, 1971.

Sellek, Yoko. "Female Foreign Migrant Workers in Japan: Working for the Yen." *Japan Forum* 8, no. 2 (1996): 159–175.

Shamoon, Deborah. "Office Sluts and Rebel Flowers: The Pleasures of Japanese Pornographic Comics for Women." In *Porn Studies*, edited by Linda Williams, 77–103. Durham, NC: Duke University Press, 2004.

Shibuya Hetamoji Fukyu Iinkai. *Gyaru moji heta moju kōshiki Book* (Primer book for girl characters and odd characters). Tokyo: Jitsugyōno Nihonsha, 2004.

Shimakawa, Karen. *National Abjection: The Asian American Body on Stage*. Durham, NC: Duke University Press, 2002.

Shinotsuka Eiko. *Josei to kazoku: Kindaika no jitsuzō* (Women and family: The real image of modernization). Tokyo: Yomiuri Shinbunsha, 1995.

Shively, Donald H. "The Social Environment of Tokugawa Kabuki." In *Studies in Kabuki: Its Acting, Music, and Historical Context*, edited by James R. Brandon, William P. Malm and Donald H. Shively, 1–62. Honolulu: University Press of Hawai'i, 1978.

Shōji Kaori. It's Always Oh, So French in the Ginza." *International Herald Tribune Online*, October 13, 2003.

Sievers, Sharon L. *Flowers in Salt: The Beginnings of Feminist Consciousness in Modern Japan*. Stanford: Stanford University Press, 1983.

Silver, Mark. "The Lies and Connivances of an Evil Woman: Early Meiji Realism and the Tale of Takanashi Oden the she-Devil." *Harvard Journal of Asiatic Studies* 63, no.1 (2003): 5–67.

Silverberg, Miriam. "The Modern Girl as Militant." In *Recreating Japanese Women, 1600–1945*, edited by Gail Lee Bernstein, 239–266. Berkeley: University of California Press, 1991.

Skov, Lise and Brian Moeran, eds. *Women, Media, and Consumption in Japan*. Honolulu: University of Hawai'i Press, 1995.

Sontag, Susan. "Notes on Camp." In *Against Interpretation*, 275–292. New York: Farrar, Straus, Giroux, 1966.

Spielvogel, Laura. *Working Out in Japan: Shaping the Female Body in Tokyo Fitness Clubs*. Durham, NC: Duke University Press, 2003.

Statistics Bureau, Management and Coordination Agency. *Jinkō tōkei sōran* (Population statistics of Japan). Tokyo: Toyō Keizai Shinpōsha, 1985.

Statler, Oliver. *The Black Ship Scroll: An Account of the Perry Expedition at Shimada in 1854*. Tokyo: John Weatherhell, 1964.

Stoller, Debbie. "Feminists Fatale." In *The Bust Guide to the New Girl Order*, edited by Marcelle Karp and Debbie Stoller, 41–47. New York: Penguin Books, 1999.

Suzuki Aya. "Nogaretsuzukeru 'josei' tachi: 'Futon' 'tandegi' no 'josei' no ichi (Fleeing women: Women's position in 'The Quilt' and 'Addiction')." *Kokugo kokubun kenkyū* 121, July (2002): 35–52.

Suzuki, Michiko. "Developing the Female Self: Same-Sex Love, Love Marriage and Maternal Love in Modern Japanese Literature, 1910–1939." Ph.D. diss., Stanford University, 2002.

Suzuki, Nobue. "Between Two Shores: Transnational Projects and Filipina Wives in/from Japan." *Women's Studies International Forum* 23, no. 4 (2000a): 431–444.

———. "Women Imagined, Women Imaging: Re/Presentations of Filipinas in Japan since the 1980s." *U.S.-Japan Women's Journal English Supplement* 19 (2000b): 142–175.

———. "Gendered Surveillance and Sexual Violence in Filipina Pre-Migration Experiences to Japan." In *Gender Politics in the Asia Pacific Region*, edited by Brenda Yeoh, Peggy Teo, and Shirlena Huang, 99–119. London: Routledge, 2002.

———. "Transgressing 'Victims': Reading Narratives of 'Filipina Brides' in Japan." *Critical Asian Studies* 35, no. 3 (2003): 399–420.

———. "Inside the Home: Power and Negotiation in Filipina-Japanese Marriages." *Women's Studies: An Interdisciplinary Journal* 33, no. 4 (2004): 481–506.

———. "Tripartite Desires: Filipina-Japanese Marriages and Fantasies of Transnational Traversal." In *Cross-Border Marriages: Gender and Mobility in Transnational Asia*, edited by Nicole Constable, 124–144. Philadelphia: University of Pennsylvania Press, 2005.

Suzuki Rumiko. *Herumesu o amaku miru to itai me ni au* (Don't take Hermès lightly or you'll be sorry). 3rd ed. Tokyo: Kōdansha, 2002.

Sykes, Plum. *Bergdorf Blondes*. New York: Hyperion, 2004.

Tamanoi, Mariko Asano. *Under the Shadow of Nationalism: Politics and Poetics of Rural Japanese Women*. Honolulu: University of Hawai'i Press, 1998.

Tanabe Seiko. *Yume haruka Yoshiya Nobuko: Akitomoshi tsukue no ue no ikusanka* (Yoshiya Nobuko's distant dreams: Autumn light and landscapes over her desk). Tokyo: Asahi Bunko, 1999. Reprint, 2002.

Tanaka Yasuo. *Nantonaku, kurisutaru* (Somehow crystal). Tokyo: Shinchō Bunko, 1981.

Tanizaki Jun'ichirō. *Naomi.* Translated by Anthony H. Chambers. New York: North Point Press, 1985.

Tayama Katai. *Futon.* Tokyo: Iwanami Shoten, 1998.

Tobin, Joseph, ed. *Re-Made in Japan: Everyday Life and Consumer Taste in a Changing Society.* New Haven: Yale University, 1992.

Todd, Rebecca. "A Butoh-full Mind: Meiko Ando dances a Japanese legend." *Eye Weekly*, an online journal, April 4, 2002: <http://www.eye.net/eye/issue/issue_04.04.02/arts/onibaba.html>

Tomida, Hiroko. *Hiratsuka Raichō and Early Japanese Feminism.* Leiden-Boston: Brill, 2004.

Tsunoda, Ryūsaku, Wm. Theodore de Bary and Donald Keene, eds., comps. Vol. 1 of *Sources of Japanese Tradition.* 2 vols. New York: Columbia University Press, 1958.

Tsushima Yūko. *Woman Running in the Mountains.* Translated by Geraldine Harcourt. New York: Pantheon Books, 1991.

Tucker, Anne Wilkes, Dana Friis-Hansen, Kaneko Ryuchi, and Joe Takeba. *The History of Japanese Photography.* New Haven: Yale University Press, 2003.

Ueno Chizuko and Nobuta Sayoko. *Kekkon teikoku: Onna no wakare michi* (Empire of marriage: Crossroads for women). Tokyo: Kōdansha, 2004.

Ury, Marian. *Tales of Times Now Past: Sixty-Two Stories from a Medieval Japanese Collection.* Ann Arbor: Center for Japanese Studies, 1979.

Viswanathan, Meera. "In Pursuit of the Yamamba: The Question of Female Resistance." In *The Woman's Hand: Gender and Theory in Japanese Women's Writing*, edited by Paul Gordon Schalow and Janet A. Walker, 76–99. Stanford: Stanford University Press, 1996.

Walker, Janet. *The Japanese Novel of the Meiji Period and the Ideal of Individualism.* Princeton: Princeton University Press, 1979.

Watanabe Jun'ichi. *A Lost Paradise.* Translated by Juliet Winters Carpenter. Tokyo, New York, London: Kodansha International, 2000.

Watanabe Masaaki. "Tayama Katai 'Futon' to 'Jogakusei daraku monogatari' (Tayama Katai's 'Futon' and 'Tales of degenerate schoolgirls')." *Gunma kenritsu joshi daigaku kokubungaku kenkyū* (1991): 13–26.

Watanabe Yayoi. "Bacherā Pātii (Bachelor party)," In *Kaama* Vol. 7 (June 2000), 39–70.

———. "Ji endo (The end)". In *Manon* 10, no. 7 (129) (July 2003): 4–103.

Wehrfritz, George and Kay Itoi. "The Luxury Bubble."*Newsweek*, Feburary 10, 2003, 34.

Wicke, Jennifer. "Through the Glass Darkly." In *Dirty Looks: Women, Pornography, Power*, edited by Pamela Church Gibson with Roma Gibson, 62–80. London: British Film Institute, 1993.

Williams, Linda. "When Women Look: A Sequel." *Senses of Cinema: An Online Journal Devoted to the Serious and Eclectic Discussion of Cinema (SoC)* (2001): unpaginated. Online at <http://www.sensesofanema.com/contents/01/15/horror-women.html>

Willis, Ellen. "Feminism, Moralism and Pornography." In *Beginning to See the Light: Pieces of a Decade*, Ellen Willis, 219–227. New York: Knopf, 1981.

Wilson, Elizabeth. "The Female Body as a Source of Horror and Insight in Post-Ashokan Indian Buddhism." In *Religious Reflections on the Human Body*, edited by Jane Marie Law, 76–99. Bloomington: Indiana University Press, 1995.

Wright, Diana E. "Female Crime and State Punishment in Early Modern Japan." *Journal of Women's History* 16, no. 3 (2004): 10–29.

Yamada Toyoko. *Burando no seiki* (Brand in twentieth century). Tokyo: Magajin Hausu, 2000.

Yamane Kazuma. *Hentai shōjo mōji* (Morphology of girls' handwriting). Tokyo: Kōdansha, 1989.

———. *Gyaru no kōzō* (The structure of the girl). Tokyo: Kōdansha, 1993.

Yanagita Kunio. "Imo no chikara (The power of women)." In *Teihon Yanagita Kunio shū*, 1–219. Tokyo: Chikuma Shobō, 1962.

Yoshida, Ruiko. *Hot Harlem Days*. Tokyo Kōdansha, 1967.

Yoshikawa Toyoko. "*Seitō* kara 'taishū shōsetsu e' no michi: Yoshiya Nobuko *Yaneura no nishojo* (The Road from *Bluestockings* to 'popular fiction': Yoshiya Nobuko's *Two Virgins in the Attic*)." In *Feminizumu hihyō e no shōtai* (An introduction to feminist criticism), edited by Iwabuchi Hiroko et al., 121–147. Tokyo: Gakugei Shorin, 1995.

Yoshitake Teruko. *Nyonin Yoshiya Nobuko* (The woman Yoshiya Nobuko). Tokyo: Bungei Shunjū, 1986.

Yoshiya Nobuko. "Haru o mukae no bundan susuharai (A spring cleaning for the literary establishment)." *Kuroshōbi* (1925): 64–65.

———. "Danpatsu oshikari no koto (Censure of the *danpatsu*)." *Kuroshōbi* 3 (March 1925): 52–56.

———. *Jidenteki joryū bundanshi* (Autobiographical history of the women's literary establishment). Tokyo: Chūō Kōronsha, 1962.

———. *Zuihitsu: Watashi no mita bijintachi* (Essays: Beauties as I have seen them). Tokyo: Yomiuri Shimbunsha, 1969.

———. "Tōsho jidai (In my days of reader submissions)." In *Yoshiya Nobuko: Sakka no jiden*, Vol. 66 (Yoshiya Nobuko: Authors' autobiographies). Tokyo: Nihon Tosho Sentā, 1998.

———. "Foxfire (Onibi)." Translated by Lawrence Rogers. *The East* 36, no. 1 (May–June 2000), 41–43.

———. *Hana monogatari* (Flower stories). Originally published in 1920. Tokyo: Kokushokan, 1995.

———. *Otome shōsetsu korekushon* (Collection of maiden fiction). Tokyo: Kokushokan, 2002–2003.

———. *Yaneura no nishojo* (Two virgins in the attic). Tokyo: Rakuyōdō, 1920. Reprint, Kokushokan, 2003.

Yuasa, Michiko. "Women in Shinto: Images Remembered." In *Religion and Women*, edited by Arvind Sharma, 93–119. Albany, NY: State University of New York Press, 1994.

Zenikkū Eigyōhonbu Kyōiku Kunrenbu, ed. *OL tabū shū* (Anthology of office lady taboos). Tokyo: Goma Seibo, 1991.

Zonana, Joyce. "The Sultan and the Slave: Feminist Orientalism and the Structure of Jane Eyre." In *Revising the Word and the World: Essays in Feminist Literary Criticism*, edited by Vévé A. Clark, Ruth-Ellen B. Joeres, and Madelon Sprengnether, 167–190. Chicago: University of Chicago Press, 1993.

Contributors

Jan Bardsley is associate professor of Japanese humanities at the University of North Carolina at Chapel Hill. She is the author of *The Bluestockings of Japan: New Women Fiction and Essays from Seitō, 1911–1916* (Ann Arbor, MI: Center for Japanese Studies, 2006). Her current book project is *Designs on Democracy: Fashion and Feminism in Postwar Japan*.

Rebecca Copeland is associate professor of Japanese literature at Washington University in St. Louis, Missouri. Her published works include *The Sound of the Wind: The Life and Works of Uno Chiyo* (University of Hawai'i Press, 1992), *Lost Leaves: Women Writers of Meiji Japan* (University of Hawai'i Press, 2000), and *The Father-Daughter Plot: Japanese Literary Women and the Law of the Father* (University of Hawai'i Press, 2001), which she coedited with Dr. Esperanza Ramirez-Christensen of University of Michigan.

Melanie Czarnecki is a lecturer in the faculty of foreign studies at Sophia University in Tokyo. She is currently completing her doctoral dissertation at Hokkaido University on the politics of gender in the writings of Hiratsuka Raichō and other Meiji and Taishō women writers.

Kelly Foreman is a lecturer in the departments of anthropology and music at Wayne State University, as well as a freelance composer and a performer of *shamisen*. She has published articles on Japanese music and geisha in Japan and authored liner notes for Japanese traditional music recordings. Her forthcoming book entitled *The Gei of Geisha: Music, Identity, and Meaning* details the relationship between geisha and music.

Sarah Frederick is assistant professor of Japanese literature at Boston University. Her first book, forthcoming from University of Hawai'i Press, is entitled *Turning Pages: Reading and Writing Women's Magazines in Interwar Japan*. She is currently working on a study of the life and works of Yoshiya Nobuko.

Hiroko Hirakawa is assistant professor of Japanese and intercultural studies at Guilford College in Greensboro, North Carolina. She has published articles on popular culture and gender in contemporary Japan. Her current book project is *Lost and Found in Translation: Japanese Women Crossing Borders*.

Gretchen I. Jones is assistant professor of Japanese literature at the University of Maryland, where she teaches modern Japanese literature and language. She is the author of "Subversive Strategies: Masochism, Gender and Power in Kono Taeko's 'Toddler-Hunting' " (*East Asia* 18: 4, 2000), and "Ladies' Comics: Japan's Not-So-Underground Market in Pornography for Women," *U.S.-Japan Women's Journal* (English Supplement, Vol. 22, 2002). She is currently completing a book called *Whip Appeal: The Aesthetics of Masochism in Modern Japanese Narrative,* forthcoming from University of Hawai'i Press.

Sharon Kinsella is research associate at the Institute of Social and Cultural Anthropology, Oxford University. She has worked on men's comics, cuteness and infantilism, otaku, corporate culture, cultural governance and cultural production, and girls' culture in contemporary Japan. A list of previous publications can be found at <http://www. kinsellaresearch.com>. The chapter "Black Faces, Witches and Racism against Girls" draws from sections of a chapter on race included in her forthcoming book *Girls As Energy: Fantasies of Rejuvenation.*

Christine Marran is assistant professor at the University of Minnesota. She teaches and has published articles on modern Japanese literature, gender, and film. Her first book, *She Had It Coming: The Poison Woman in Japanese Modernity* is forthcoming from the University of Minnesota Press.

Katherine Mezur is assistant professor of dance at Mills College, Oakland. She is the author of *Beautiful Boys: Performing Female-likeness on the Kabuki Stage* (Palgrave Macmillan, 2005). She is completing her second book, *Cute Mutant Girls: Remapping the Female Body in Contemporary Japanese Performance and Media,* and her "media performance installation" Skin, which investigates the politics of screen and kinaesthetic perception.

Laura Miller is associate professor of anthropology at Loyola University Chicago. She has published widely in the fields of linguistic anthropology, popular culture, and gender studies. Her book, *Beauty Up: Exploring Contemporary Japanese Body Aesthetics,* is forthcoming from the University of California Press.

Miriam Silverberg is professor of Japanese history at UCLA where she teaches comparative thought and culture. She is the author of *Changing Song: The Marxist Manifestos of Nakano Shigeharu* (Princeton University Press, 1990; winner of the John King Fairbank Award) and *Erotic Grotesque Nonsense: Japanese Modern Times* (University of California Press, forthcoming) and of recent articles such as "Transnational Wives' Tales: Yonsama as Postcolonial Ghost and the Revenge of the *Obatarians*" (*Contemporary Women's History in Asia*) and "War Responsibility Revisited: Auschwitz in Japan" (*Journal of Asian and Pacific Studies*). She is currently studying the gendering of post Pacific War Japanese wartime sentiment.

Nobue Suzuki, professor of anthropology at Nagasaki Wesleyan University, is the coeditor of *Men and Masculinities in Contemporary Japan* (Routledge, 2003, with James E. Roberson) and is working on a book tentatively entitled, *Battlefields of Affection: Romanscapes and Filipino-Japanese Families.* Her recent publications include, "Inside the Home: Power and Negotiation in Filipina-Japanese Marriages" (*Women's Studies,* 2004) and "Tripartite Desires: Filipina-Japanese Marriages and Fantasies of Transnational Traversal" (in Nicole Constable, ed., *Cross-Border Marriages: Gender and Mobility in Transnational Asia,* University of Pennsylvania Press, 2005).

Index

Boldface locators indicate illustrations; locators followed by n indicate footnotes.